'Thanks to Michio Kaku, you don't have to be an Einstein to understand Einstein. *Einstein's Cosmos* weaves together Einstein's life and science the way Einstein himself wove together time and space'   Dr Ken Croswell, author of
Magnificent Universe and Magnificent Mars

'Michio Kaku beautifully interweaves the story of Albert Einstein's career with Einstein's stunning insights into cosmology, from relativity and black holes to dark energy and the search for a unified theory of all the forces in the universe'
Donald Goldsmith, author of
The Runaway Universe and *Connecting with the Cosmos*

'Kaku is a skilful and genial populariser . . . While this is fully a biography, succinctly revisiting Einstein's difficult childhood and unpromising early career, his two marriages and his emergence as a 20th-century cultural icon, the guiding thread is undoubtedly the evolution of his ideas'   *Sunday Telegraph*

'One of the most sympathetic and also scientifically interesting biographies of Einstein to ever appear in print . . . a fascinating and easy read'   *Focus*

'A memorable book . . . a gem'   *Good Book Guide*

'It's started – the worldwide frenzy to mark next year's centenary of Einstein's "miraculous year" . . . If you want to know what all the fuss will be about, read Michio Kaku's authoritative offering'
New Scientist

One of the most prominent and respected scientists today, Michio Kaku holds the Henry Semat Professorship in Theoretical Physics at the Graduate Center of the City University of New York and the City College of New York. He is the co-founder of string field theory. He is the author of *Hyperspace, Beyond Einstein* (with Jennifer Trainer), and *Visions: How Science Will Revolutionize the 21st Century*, as well as numerous PhD-level textbooks that are required reading at many of the world's leading universities. His weekly radio show, 'Explorations', can be heard on stations across America, and he has frequently appeared on television talk shows and BBC and Public Television science specials. He is the host of a four-part BBC TV special on the nature of time. Professor Kaku lives in New York City. His website can be visited at www.mkaku.org.

EINSTEIN'S COSMOS

How Albert Einstein's Vision
Transformed Our Understanding
of Space and Time

MICHIO KAKU

PHOENIX

A PHOENIX PAPERBACK

First published in Great Britain in 2004
by Weidenfeld & Nicolson
This paperback edition published in 2005
by Phoenix,
an imprint of Orion Books Ltd,
Orion House, 5 Upper St Martin's Lane,
London WC2H 9EA

1 3 5 7 9 10 8 6 4 2

A CIP catalogue record for this book
is available from the British Library.

ISBN 0 75381 904 X
EAN 9 780753 819043

Printed and bound in Great Britain
Clays Ltd, St Ives plc

www.orionbooks.co.uk

This book is dedicated to Michelle and Alyson.

Contents

A New Look at the Legacy
of Albert Einstein

Genius. Absent-minded professor. The father of relativity. The mythical figure of Albert Einstein—hair flaming in the wind, sockless, wearing an oversized sweatshirt, puffing on his pipe, oblivious to his surroundings—is etched indelibly on our minds. "A pop icon on a par with Elvis Presley and Marilyn Monroe, he stares enigmatically from postcards, magazine covers, T-shirts, and larger-than-life posters. A Beverly Hills agent markets his image for television commercials. He would have hated it all," writes biographer Denis Brian.

Einstein is among the greatest scientists of all time, a towering figure who ranks alongside Isaac Newton for his contributions. Not surprisingly, *Time* magazine voted him the Person of the Century. Many historians have placed him among the hundred most influential people of the last thousand years.

Given his place in history, there are several reasons for trying to make a fresh new effort to re-examine his life. First, his theories are so deep and profound that the predictions he made decades ago are still dominating the headlines, so it is vital that we try to understand the roots of these theories. As a new generation of instruments that were inconceivable in the 1920s (e.g., satellites, lasers, supercomputers, nanotechnology, gravity wave detectors) probe the outer reaches of the cosmos and the interior of the atom, Einstein's predictions are winning Nobel Prizes for other scientists. Even the crumbs off Einstein's table are opening up new vistas for science. The 1993 Nobel Prize, for example, went to two physicists who indirectly confirmed the existence of gravity waves, predicted by Einstein in 1916, by analyzing the motion of double neutron stars in the heavens. Also, the 2001 Nobel Prize went to three physicists who confirmed the existence of Bose-Einstein condensates, a new state of matter existing near absolute zero that Einstein predicted in 1924.

Other predictions are now being verified. Black holes, once considered a bizarre aspect of Einstein's theory, have now been identified by the Hubble Space Telescope and the Very Large Array Radio Telescope. Einstein rings and Einstein lenses not only have been confirmed but also are key tools astronomers use to measure invisible objects in outer space.

Even Einstein's "mistakes" are being recognized as profound contributions to our knowledge of the universe. In 2001, astronomers found convincing evidence that the "cosmological constant," thought to be Einstein's greatest blunder, actually contains the largest concentration of energy in the universe and will determine the ultimate fate of the cosmos itself. So experimentally, there has been a "renaissance" in Einstein's legacy as more evidence piles up verifying his predictions.

Second, physicists are re-evaluating his legacy and especially his thinking process. While recent biographies have minutely examined his private life for clues to the origins of his theories, physicists are becoming increasingly aware that Einstein's theories are based not so much on arcane mathematics (let alone his love life!) but simple and elegant physical pictures. Einstein would often comment that if a new theory was not based on a physical image simple enough for a child to understand, it was probably worthless.

In this book, therefore, these pictures, these products of Einstein's scientific imagination, become a formal organizing principle around which his thinking process and his greatest achievements are described.

Part I uses the picture that Einstein first thought of when he was sixteen years old: what a light beam would look like if he could race alongside it. This picture, in turn, was probably inspired by a children's book that he read. By visualizing what happens if he were to race a light beam, Einstein isolated the key contradiction between the two great theories of the time, Newton's theory of forces and Maxwell's theory of fields and light. In the process of resolving this paradox, he knew that one of these two great theories—Newton's, as it turns out—must fall. In some sense, all of special relativity (which would eventually unlock the secret of the stars and nuclear energy) is contained in this picture.

In Part II, we are introduced to another picture: Einstein imagined planets as marbles rolling around a curved surface centered at the sun, as an illustration of the idea that gravity originates from the bending of space and time. By replacing the forces of Newton with the curvature of a smooth surface, Einstein gave an entirely fresh, revolutionary picture of gravity. In this new framework, the "forces" of Newton were an illusion caused by the bending of space itself. The consequences of this simple picture would eventually give us black holes, the big bang, and the ultimate fate of the universe itself.

Part III doesn't have a picture—this section is more about the failure to come up with an image guiding his "unified field theory," one that would have given Einstein a way to formulate the crowning achievement of two thousand years of investigation into the laws of matter and energy. Einstein's intuition began to falter, as almost nothing was known in his time about the forces that governed the nucleus and subatomic particles.

This unfinished unified field theory and his thirty-year search for a "theory of everything" was by no means a failure—although this has been recognized only recently. His contemporaries saw it as a fool's chase. The physicist and Einstein biographer Abraham Pais lamented, "In the remaining 30 years of his life he remained active in research but his fame would be undiminished, if not enhanced, had he gone fishing instead." In other words, his legacy might have been even greater if he had left physics in 1925 rather than 1955.

In the last decade, however, with the coming of a new theory called "superstring theory" or "M-theory," physicists have been re-evaluating Einstein's later work and his legacy, as the search for the unified field theory has assumed center stage in the world of physics. The race to attain the theory of everything has become the ultimate goal of a whole generation of young, ambitious scientists. Unification, once thought to be the final burial ground for the careers of aging physicists, is now the dominant theme in theoretical physics.

In this book, I hope to give a new, refreshing look at the pioneering work of Einstein, perhaps even a more accurate portrayal of his enduring legacy as seen from the vantage point of simple physical pictures. His insights, in turn, have fueled the current generation of revolutionary new experiments being conducted in outer space and in advanced physics laboratories and are driving the intense search to fulfill his most cherished dream, a theory of everything. This is the approach to his life and his work that I think he would have liked the best.

Acknowledgments

I would like to thank the hospitality of the staff at Princeton University Library, where some of the research for this book was carried out. The library contains copies of all of Einstein's manuscripts and original materials. I would also like to thank Professors V. P. Nair and Daniel Greenberger of City College of New York for reading the manuscript and making helpful and critical comments. In addition, conversations with Fred Jerome, who obtained Einstein's voluminous FBI file, were very useful. I am also grateful to Edwin Barber for his support and encouragement, and to Jesse Cohen for making invaluable editorial comments and changes that have greatly strengthened the manuscript and given it focus. I am also deeply indebted to Stuart Krichevsky, who has represented many of my books on science for all these years.

EINSTEIN'S
COSMOS

PART I

FIRST PICTURE

Racing a Light Beam

CHAPTER 1

Physics before Einstein

Ajournalist once asked Albert Einstein, the greatest scientific genius since Isaac Newton, to explain his formula for success. The great thinker thought for a second and then replied, "If A is success, I should say the formula is $A = X + Y + Z$, X being work and Y being play."

And what is Z, asked the journalist?

"Keeping your mouth shut," he replied.

What physicists, kings and queens, and the public found endearing was his humanity, his generosity, and his wit, whether he was championing the cause of world peace or probing the mysteries of the universe.

Even children would flock to see the grand old man of physics walk the streets of Princeton, and he would return the favor by wiggling his ears back at them. Einstein liked to chat with a particular five-year-old boy who would accompany the great thinker on his walks to the Institute for Advanced Study. One day while they were strolling, Einstein suddenly burst out in laughter. When the boy's mother asked him what they talked about, her son replied, "I asked Einstein if he had gone to the bathroom today." The mother was horrified, but then Einstein replied, "I'm glad to have someone ask me a question I can answer."

As physicist Jeremy Bernstein once said, "Everyone who had real contact with Einstein came away with an overwhelming sense of the nobility of the man. The phrase that recurs again and again is his 'humanity,' . . . the simple, lovable quality of his character."

Einstein, who was equally gracious to beggars, children, and royalty alike, was also generous to his predecessors in the illustrious pantheon of science. Although scientists, like all creative individuals, can be notoriously jealous of their rivals and engage in petty squabbles, Einstein went out of his way to trace the origins of the ideas he pioneered back to the giants of physics, including Isaac Newton and James Clerk Maxwell, pictures of whom were prominently displayed on his desk and walls. In fact, the work of Newton on mechanics and gravity and of Maxwell on light formed the two pillars of science at the turn of the twentieth century. Remarkably, almost the sum total of all physical knowledge at that time was embodied in the achievements of these two physicists.

It's easy to forget that before Newton, the motion of objects on Earth and in the heavens was almost totally unexplained, with many believing that our fates were determined by the malevolent designs of spirits and demons. Witchcraft, sorcery, and superstition were heatedly debated even at the most learned centers of learning in Europe. Science as we know it did not exist.

Greek philosophers and Christian theologians, in particular, wrote that objects moved because they acted out of human-like desires and emotions. To the followers of Aristotle, objects in motion eventually slowed down because they got "tired." Objects fell to the floor because they "longed" to be united with the earth, they wrote.

The man who would introduce order into this chaotic world of spirits was in a sense the opposite of Einstein in temperament and personality. While Einstein was always generous with his time and quick with a one-liner to delight the press, Newton was notoriously reclusive, with a tendency toward paranoia. Deeply suspicious of others, he had bitter, long-standing feuds with other scientists over priority. His reticence was legendary: when he was a member of the British Parliament during the 1689–90 session, the only recorded incident of his speaking before the august body was when he felt a draft and asked an usher to close the window. According to biographer Richard S. Westfall, Newton was a "tortured man, an extremely neurotic personality who teetered always, at least through middle age, on the verge of breakdown."

But in matters of science, Newton and Einstein were true masters, sharing many key characteristics. Both could obsessively spend weeks and months in intense concentration to the point of physical exhaustion and collapse. And both had the ability to visualize in a simple picture the secrets of the universe.

In 1666, when Newton was twenty-three years old, he banished the spirits that haunted the Aristotelian world by introducing a new mechanics based on *forces*. Newton proposed three laws of motion in which objects moved because they were being pushed or pulled by forces that could be accurately measured and expressed by simple equations. Instead of speculating on the desires of objects as they moved, Newton could compute the trajectory of everything from falling leaves, soaring rockets, cannonballs, and clouds by adding up the forces acting on them. This was not merely an academic question, because it helped to lay the foundation for the Industrial Revolution, where the power of steam engines driving huge locomotives and ships created new empires. Bridges, dams, and towering skyscrapers could now be built with great confidence, since the stresses on every brick or beam could be computed. So great was the victory of Newton's theory of forces that he was justly lionized during his lifetime, prompting Alexander Pope to acclaim:

> *Nature and Nature's laws lay hid in night,*
> *God said, Let Newton be! and all was light.*

Newton applied his theory of forces to the universe itself by proposing a new theory of gravity. He liked to tell the story of how he returned to the family estate in Woolsthorpe in Lincolnshire after the black plague forced the closing of Cambridge University. One day, as he saw an apple fall off a tree on his estate, he asked himself the fateful question: if an apple falls, then does the moon also fall? Can the gravitational force acting on an apple on Earth be the same force that guides the motion of heavenly bodies? This was heresy, since the planets were supposed to lie on fixed spheres that obeyed perfect, celestial laws, in contrast to the laws of sin and redemption that governed the wicked ways of humanity.

In a flash of insight, Newton realized he could unify both earthly and heavenly physics into one picture. The force that pulled an apple to the ground must be the same force that reached out to the moon and guided its path. He stumbled upon a new vision of gravity. He imagined himself sitting on a mountaintop throwing a rock. By throwing the rock faster and faster, he realized that he could throw it farther and farther. But then he made the fateful leap: what happens if you throw the rock so fast that it never returns? He realized that a rock, falling continually under gravity, would not hit the earth but would circle around it, eventually returning to its owner and hitting him on the back of his head. In this new view, he replaced the rock with the moon, which was constantly falling but never hit the ground because, like the rock, it moved completely around the earth in a circular orbit. The moon was not resting on a celestial sphere, as the church thought, but was continually in free fall like a rock or apple, guided by the force of gravity. This was the first explanation of the motion of the solar system.

Two decades later, in 1682, all of London was terrified and amazed by a brilliant comet that was lighting up the night sky. Newton carefully tracked the motion of the comet with a reflecting telescope (one of his inventions) and found that its motion fit his equations perfectly if it was assumed to be in free fall and acted on by gravity. With the amateur astronomer Edmund Halley, he could predict precisely when the comet (later known as Halley's comet) would return, the first prediction made on the motion of comets. The laws of gravity that Newton used to calculate the motion of Halley's comet and the moon are the same ones NASA uses today to guide its space probes with breathtaking accuracy past Uranus and Neptune.

According to Newton, these forces act instantaneously. For example, if the sun were to suddenly disappear, Newton believed the earth would be instantly thrown out of its orbit and would freeze in deep space. Everyone in the universe would know that the sun had just disappeared at that precise instant of time. Thus, it's possible to synchronize all watches so they beat uniformly anywhere in the universe. A second on Earth has the same length as a second on Mars and Jupiter. Like time, space is also absolute. Meter sticks on Earth have the same length as meter sticks on Mars and Jupiter. Meter sticks do not change in length anywhere in the universe. Seconds and meters are therefore the same no matter where we journey in space.

Newton thus based his ideas on the commonsense notion of *absolute space and time*. To Newton, space and time formed an absolute reference frame against which we can judge the motion of all objects. If we are traveling on a train, for example, we believe that the train is moving and the earth is still. However, after gazing at the trees passing our windows, we can speculate that perhaps the train is actually at rest, and the trees are being sent past our windows. Since everything in the train seems motionless, we can ask the question, which is really moving, the train or the trees? To Newton, this absolute reference frame could determine the answer.

Newton's laws remained the foundation for physics for nearly two centuries. Then, in the late nineteenth century, as new inventions such as the telegraph and the light bulb were revolutionizing the great cities of Europe, the study of electricity brought about a whole new concept in science. To explain the mysterious forces of electricity and magnetism, James Clerk Maxwell, a Scottish physicist at Cambridge University working in the 1860s, developed a theory of light not based on Newtonian forces, but on a new concept called *fields*. Einstein wrote that the field concept "is the most profound and the most fruitful that physics has experienced since Newton."

These fields can be visualized by sprinkling iron filings over a sheet of paper. Place a magnet beneath the sheet of paper, and the filings will magically rearrange themselves into a spider web–like pattern, with lines spreading from the North Pole to the South Pole. Surrounding any magnet, therefore, is a magnetic field, an invisible array of lines of force penetrating all of space.

Electricity creates fields as well. At science fairs, children laugh when their hairs stand on end as they touch a source of static electricity. The hairs trace out the invisible electric field lines emanating from the source.

These fields, however, are quite different from the forces introduced by Newton. Forces, said Newton, act instantly over all space, so that a disturbance in one part of the universe would be felt instantly throughout all the universe. Maxwell's brilliant observation was that magnetic and electric effects do not travel instantaneously, like Newtonian forces, but take time and move at a definite velocity. His biographer Martin Goldman writes, "The idea of the *time* of magnetic action . . . seems to have struck Maxwell like a bolt out of the blue." Maxwell showed, for example, that if one shook a magnet, it would take time for nearby iron filings to move.

Imagine a spider web vibrating in the wind. A disturbance like the wind on one part of the web causes a ripple that spreads throughout the entire web. Fields and spider webs, unlike forces, allow for vibrations that travel at a definite speed. Maxwell then set out to calculate the speed of these magnetic and electric effects. In one of the greatest breakthroughs of the nineteenth century, he used this idea to solve the mystery of light.

Maxwell knew from the earlier work of Michael Faraday and others that a moving magnetic field can create an electric field, and vice versa. The generators and motors that electrify our world are direct consequences of this dialectic. (This principle is used to light up our homes. For example, in a dam, falling water spins a wheel, which in turn spins a magnet. The moving magnetic field pushes the electrons in a wire, which then travel in a high-voltage wire to the wall sockets in our living rooms. Similarly, in an electric vacuum cleaner, the electricity flowing from our wall sockets creates a magnetic field that forces the blades of the motor to spin.)

The genius of Maxwell was to put the two effects together. If a changing magnetic field can create an electric field and vice versa, then perhaps both of them can form a cyclical motion, with electric fields and magnetic fields continually feeding off each other and turning into each other. Maxwell quickly realized that this cyclical pattern would create a moving train of electric and magnetic fields, all vibrating in unison, each turning into the other in a never-ending wave. Then he calculated the speed of this wave.

To his astonishment, he found that it was the speed of light. Further, in perhaps the most revolutionary statement of the nineteenth century, he claimed that this *was* light. Maxwell then announced prophetically to his colleagues, "We can scarcely avoid the conclusion that *light consists of the transverse undulations of the same medium which is the cause of electric and magnetic phenomenon.*" After puzzling over the nature of light for millennia, scientists finally understood its deepest secrets. Unlike Newton's forces, which were instantaneous, these fields traveled at a definite speed: the speed of light.

Maxwell's work was codified in eight difficult partial differential equations (known as "Maxwell's equations"), which every electrical engineer and physicist has had to memorize for the past century and a half. (Today, one can buy a T-shirt containing all eight equations in their full glory. The T-shirt prefaces the equations by stating, "In the beginning, God said . . . ," and ends by saying, " . . . and there was light.")

By the end of the nineteenth century, so great were the experimental successes of Newton and Maxwell that some physicists confidently predicted that these two great pillars of science had answered all the basic questions of the universe. When Max Planck (founder of the quantum theory) asked his advisor about becoming a physicist, he was told to switch fields because physics was basically finished. There was nothing really new to be discovered, he was told. These thoughts were echoed by the great nineteenth-century physicist Lord Kelvin, who proclaimed that physics was essentially complete, except for a few minor "clouds" on the horizon that could not be explained.

But the deficiencies of the Newtonian world were becoming more and more glaring each year. Discoveries like Marie Curie's isolation of radium and radioactivity were rocking the world of science and catching the public imagination. Even a few ounces of this rare, luminous substance could somehow light up a darkened room. She also showed that seemingly unlimited quantities of energy could come from an unknown source deep inside the atom, in defiance of the law of conservation of energy, which states that energy cannot be created or destroyed. These small "clouds," however, would soon spawn the great twin revolutions of the twentieth century, relativity and the quantum theory.

But what seemed most embarrassing was that any attempt to merge Newtonian mechanics with Maxwell's theory failed. Maxwell's theory confirmed the fact that light was a wave, but this left open the question, what is waving? Scientists knew that light can travel in a vacuum (in fact traveling millions of light-years from distant stars through the vacuum of outer space), but since a vacuum by definition is "nothing," this left the paradox that nothing was waving!

Newtonian physicists tried to answer this question by postulating that light consisted of waves vibrating in an invisible "aether," a stationary gas that filled up the universe. The aether was supposed to be the absolute reference frame upon which one could measure all velocities. A skeptic might say that since the earth moved around the sun, and the sun moved around the galaxy, then it was impossible to tell which was really moving. Newtonian physicists answered this by stating that the solar system was moving with respect to the stationary aether, so one could determine which was really moving.

However, the aether began to assume more and more magical and bizarre properties. For example, physicists knew that waves travel faster in a denser medium. Thus, sound vibrations can travel faster in water than in air. However, since light traveled at a fantastic velocity (186,000 miles per second), it meant that the aether must be incredibly dense to transmit light. But how could this be, since the aether was also supposed to be lighter than air? With time, the aether became almost a mystical substance: it was absolutely stationary, weightless, invisible, with zero viscosity, yet stronger than steel and undetectable by any instrument.

By 1900, the deficiencies of Newtonian mechanics were becoming harder and harder to explain. The world was now ready for a revolution, but who would lead it? Although other physicists were well aware of the holes in the aether theory, they timidly tried to patch them up within a Newtonian framework. Einstein, with nothing to lose, was able to strike at the heart of the problem: *that Newton's forces and Maxwell's fields were incompatible. One of the two pillars of science must fall.* When one of these pillars finally fell, it would overturn more than two hundred years of physics and would revolutionize the way we view the universe and reality itself. Newtonian physics would be toppled by Einstein with a picture that even a child could understand.

CHAPTER 2

The Early Years

T he man who would forever reshape our conception of the universe was born on March 14, 1879, in the small city of Ulm, Germany. Hermann and Pauline Koch Einstein were distressed that their son's head was misshapen, and prayed that he was not mentally damaged.

Einstein's parents were middle-class secularized Jews struggling to provide for their growing family. Pauline was the daughter of a relatively wealthy man: her father, Julius Derzbacher (who changed his name to Koch), established his fortune by leaving his job as a baker and entering the grain trade. Pauline was the cultured one in the Einstein family, insisting that her children take up music and starting young Albert on his lifelong love affair with the violin. Hermann Einstein, in contrast to his father-in-law, had a lackluster business career, originally starting in the featherbed business. His brother Jakob convinced him to switch to the new electrochemical industry. The inventions of Faraday, Maxwell, and Thomas Edison, all of which harnessed the power of electricity, were now lighting up cities around the world, and Hermann saw a future building dynamos and electric lighting. The business would prove precarious, however, leading to periodic financial crises and bankruptcies and forcing the family to relocate several times during Albert's childhood, including to Munich a year after his birth.

The young Einstein was late in learning how to speak, so late that his parents feared that he might be retarded. But when he finally did speak, he did so in complete sentences. Still, even as a nine-year-old, he could not talk very well. His only sibling was his sister Maja, two years younger than Albert. (At first, young Albert was puzzled by the new presence in the household. One of his first phrases was, "But where are the wheels?") Being the younger sister to Albert was no joy, since he had a nasty tendency to throw objects at her head. She would later lament, "A sound skull is needed to be the sister of a thinker."

Contrary to myth, Einstein was a good student in school, but he was only good in the areas he cared about, such as mathematics and science. The German school system encouraged students to give short answers based on rote memorization—otherwise, they might be punished by painful slaps to the knuckles. Young Albert, however, spoke slowly, hesitantly, choosing his words carefully. He was far from being the perfect student, chafing under a suffocating, authoritative system that crushed creativity and imagination, replacing them with mind-numbing drills. When his father asked the headmaster what profession young Albert should pursue, he replied, "It doesn't matter; he'll never make a success of anything."

Einstein's demeanor established itself early. He was dreamy, often lost in thought or reading. His classmates used to taunt him by calling him *Biedermeier*, which translates loosely as "nerd." A friend would remember, "Classmates regarded Albert as a freak because he showed no interest in sports. Teachers thought him dull-witted because of his failure to learn by rote and his strange behavior." At age ten, Albert entered the Luitpold Gymnasium in Munich, where his most excruciating ordeal was learning classical Greek. He would sit in his chair, smiling blankly to hide his boredom. At one point his seventh-grade Greek teacher, Herr Joseph Degenhart, told Albert to his face that it would be better if he simply were not there. When Einstein protested that he did nothing wrong, the teacher replied bluntly, "Yes, that is true. But you sit there in the back row and smile, and that violates the feeling of respect which a teacher needs from his class."

Even decades later, Einstein would bitterly nurse the scars left by the authoritarian methods of the day: "It is, in fact, nothing short of a miracle that the modern methods of instruction have not yet entirely strangled the holy curiosity of inquiry; for this delicate little plant, aside from stimulations, stands mainly in need of freedom."

Einstein's interest in science started early, beginning with his encounter with magnetism, which he called his "first miracle." He was given a compass by his father and was endlessly fascinated by the fact that invisible forces could make objects move. He fondly remembered, "A wonder of such nature I experienced as a child of 4 or 5 years, when my father showed me a compass needle. . . . I can still remember . . . that this experience made a deep and lasting impression upon me. Something deeply hidden had to be behind things."

When he was about eleven, however, his life took an unexpected turn: he became devoutly religious. A distant relative would come by to tutor Albert in the Jewish faith, and he latched onto it in a surprisingly enthusiastic, almost fanatical way. He refused to eat pork and even composed several songs in praise of God, which he sang on his way to school. This period of intense religious fervor did not last long, however. The further he delved into religious lore and doctrine, the more he realized that the worlds of science and religion collided, with many of the miracles found in religious texts violating the laws of science. "Through the reading of popular books I soon reached the conviction that much in the stories of the Bible could not be true," he concluded.

Just as abruptly as he picked religion up, he dropped it. His religious phase, however, would have a profound effect on his later views. His reversal represented his first rejection of unthinking authority, one of the lifelong hallmarks of his personality. Never again would Einstein unquestioningly accept authority figures as the final word. Although he concluded that one could not reconcile the religious lore found in the Bible with science, he also decided that the universe contained whole realms that were just beyond the reach of science, and that one should have profound appreciation for the limitations of science and human thought.

His early interest in compasses, science, and religion, however, might have withered had young Albert not found a caring mentor to hone his ideas. In 1889, a poor Polish medical student named Max Talmud was studying in Munich and ate weekly dinners at the Einstein house. It was Talmud who introduced Einstein to the wonders of science beyond the dry, rote memorization of his classes. Years later, Talmud would fondly write, "In all these years I never saw him reading any light literature. Nor did I ever see him in the company of school mates or other boys of his age. His only diversion was music, he already played Mozart and Beethoven sonatas, accompanied by his mother." Talmud gave Einstein a book on geometry, which he devoured day and night. Einstein called this his "second miracle." He would write, "At the age of 12, I experienced a second wonder of a totally different nature: in a little book with Euclidean plane geometry." He called it his "holy geometry book," which he treated as his new Bible.

Here at last, Einstein made contact with the realm of pure thought. Without expensive laboratories or equipment, he could explore universal truth, limited only by the power of the human mind. Mathematics, observed his sister Maja, became an endless source of pleasure to Albert, especially if intriguing puzzles and mysteries were involved. He bragged to his sister that he had found an independent proof of the Pythagorean theorem about right triangles.

Einstein's mathematical readings did not stop there; eventually he taught himself calculus, surprising his tutor. Talmud would admit, "Soon the flight of his mathematical genius was so high that I could no longer follow. . . . Thereafter, philosophy was often the subject of our conversations. I recommended that he read Kant." Talmud's exposure of young Albert to the world of Immanuel Kant and his *Critique of Pure Reason* nourished Einstein's lifelong interest in philosophy. He began to ponder the eternal questions facing all philosophers, such as the origin of ethics, the existence of God, and the nature of wars. Kant, in particular, held unorthodox views, even casting doubt about the existence of God. He poked fun at the pompous world of classical philosophy, where "there is usually a great deal of wind." (Or, as the Roman orator Cicero once said, "There is nothing so absurd that it has not been said by a philosopher.") Kant also wrote that world government was the way to end wars, a position that Einstein would hold for the rest of his life. At one point, Einstein was so moved by the musings of Kant that he even considered becoming a philosopher. His father, who wanted a more practical profession for his son, dismissed this as "philosophical nonsense."

Fortunately, because of his father's electrochemical business, there were plenty of electric dynamos, motors, and gadgets lying around the factory to nourish his curiosity and stimulate his interest in science. (Hermann Einstein was working to get the contract for an ambitious project with his brother Jakob, the electrification of the city center of Munich. Hermann dreamed of being at the forefront of this historic undertaking. If he landed the project, it would mean financial stability as well as a large expansion of his electric factory.)

Being surrounded by huge electromagnetic contraptions no doubt awakened in Albert an intuitive understanding of electricity and magnetism. In particular, it probably sharpened his remarkable ability to develop graphic, physical pictures that would describe the laws of nature with uncanny accuracy. While other scientists often buried their heads in obscure mathematics, Einstein saw the laws of physics as clear as simple images. Perhaps this keen ability dates from this happy period of time, when he could simply look at the gadgets surrounding his father's factory and ponder the laws of electricity and magnetism. This trait, the ability to see everything in terms of physical pictures, would mark one of Einstein's great characteristics as a physicist.

At age fifteen, Einstein's education was disrupted by the family's periodic financial problems. Hermann, generous to a fault, was always helping those in financial trouble; he wasn't tough-minded like most successful businessmen. (Albert would later inherit this same generosity of spirit.) His company, failing to land the contract to light up Munich, went bankrupt. Pauline's wealthy family, now living in Genoa, Italy, offered to come to Hermann's aid by backing a new company. There was a catch, however. They insisted that he move his family to Italy (in part so they could keep a tight rein on his freewheeling, overgenerous impulses). The family moved to Milan, close to a new factory in Pavia. Not wanting to further interrupt his son's education, Hermann left Albert with some distant relatives in Munich.

All alone, Albert was miserable, trapped in a boarding school he hated and facing military duty in the dreaded Prussian army. His teachers disliked him, and the feeling was mutual. He was apparently about to be expelled from school. On an impulse, Einstein decided to reunite with his family. He arranged for his family doctor to write a medical note excusing him from school, stating that he might suffer a breakdown unless he rejoined his family. He then made the solo journey to Italy, eventually winding up totally unexpected on his parents' doorstep.

Hermann and Pauline were perplexed about what to do with their son, a draft-dodging, high school dropout with no skills, no profession, and no future. He would get into long arguments with his father, who wanted him to pursue a practical profession like electrical engineering, while Albert preferred to talk about being a philosopher. Eventually, they compromised, and Albert declared he would attend the famed Zurich Polytechnic Institute in Switzerland, even though he was two years younger than most students taking the entrance exam. One advantage was that the Polytechnic did not require a high school diploma, just a passing grade on its tough entrance examination.

Unfortunately, Einstein flunked the entrance exam. He failed the French, chemistry, and biology portions, but he did so exceptionally well in the math and physics sections that he impressed Albin Herzog, the principal, who promised to admit him the following year, without Albert having to take the dreaded exam again. The head of the physics department, Heinrich Weber, even offered to allow Einstein to audit his physics classes when he was in Zurich. Herzog recommended that Einstein spend the interim year attending a high school in Aarau, just thirty minutes west of Zurich. There Albert became a lodger at the house of the high school's director, Jost Winteler, establishing a lifelong friendship between the Einstein family and the Wintelers. (In fact, Maja would later marry Winteler's son, Paul, and Einstein's friend Michele Besso would marry the eldest daughter, Anna.)

Einstein enjoyed the relaxed, liberal atmosphere of the school. Here, he was relatively free of the oppressive, authoritarian rules of the German system. He enjoyed the generosity of the Swiss, who cherished tolerance and independence of spirit. Einstein would fondly recall, "I love the Swiss because, by and large, they are more humane than the other people among whom I have lived." Remembering all the bad memories of his years in German schools, he also decided to renounce his German citizenship, a surprising step for a mere teenager. He would remain stateless for five years (until he eventually became a Swiss citizen).

Albert, flourishing in this freer atmosphere, began to shed his image as a shy, nervous, withdrawn loner, to become outgoing and gregarious, someone who was easy to talk to and who made loyal friends. Maja, in particular, began to notice a new change in her older brother as he blossomed into a mature, independent thinker. Einstein's personality would pass through several distinct phases throughout his life, the first being his bookish, withdrawn, introverted phase. In Italy and especially Switzerland, he was entering his second phase: something of an impudent, cocky, sure-of-himself bohemian, always full of clever quips. He could make people howl with his puns. Nothing would please him more than telling a silly joke that would send his friends rolling in helpless laughter.

Some called him the "cheeky Swabian." One fellow student, Hans Byland, captured Einstein's emerging personality: "Whoever approached him was captivated by his superior personality. A mocking trait around the fleshy mouth with its protruding lower lip did not encourage the philistine to tangle with him. Unconfined by conventional restrictions, he confronted the world spirit as a laughing philosopher, and his witty sarcasm mercilessly castigated all vanity and artificiality."

By all accounts, this "laughing philosopher" was also growing up to be popular with the girls. He was a wisecracking flirt, but girls also found him sensitive, easy to confide in, and sympathetic. One friend asked him for advice in love concerning her boyfriend. Another asked him to sign her autograph book, in which he inscribed a piece of silly doggerel. His violin playing also endeared him to many and put him in demand at dinner parties. Letters from that period show that he was quite popular with women's groups who needed strings to accompany the piano. "Many a young or elderly woman was enchanted not only by his violin playing, but also by his appearance, which suggested a passionate Latin virtuoso rather than a stolid student of the sciences," wrote biographer Albrecht Folsing.

One girl in particular captured his attention. Only sixteen,

Einstein fell passionately in love with one of Jost Winteler's daughters, Marie, who was two years older. (In fact, all the key women in his life would be older than he, a tendency also shared by both his sons.) Kind, sensitive, talented, Marie wished to become a teacher like her father. Albert and Marie took long walks together, often bird watching, a favorite hobby of the Winteler family. He also accompanied her with his violin while she played the piano.

Albert confessed to her his true love: "Beloved sweetheart... I have now, my angel, had to learn the full meaning of nostalgia and longing. But love gives much more happiness than longing gives pain. I only now realize how indispensable my dear little sunshine has become to my happiness." Marie returned Albert's affections and even wrote to Einstein's mother, who wrote back approvingly. The Wintelers and the Einsteins, in fact, half expected to see a wedding announcement from the two love-birds. Marie, however, felt a bit inadequate when speaking about science with her sweetheart, and thought this could be a problem in a relationship with such an intense, focused boyfriend. She realized that she would have to compete for Einstein's affections with his first true love: physics.

What consumed Einstein's attentions was not only his growing affection for Marie but also a fascination with the mysteries of light and electricity. In the summer of 1895, he wrote an independent essay about light and the aether, entitled "An Investigation of the State of the Aether in a Magnetic Field," which he sent to his favorite uncle, Caesar Koch, in Belgium. Only five pages long, it was his very first scientific paper, arguing that the mysterious force called magnetism that mesmerized him as a child could be viewed as some kind of disturbance in the aether. Years earlier, Talmud had introduced Einstein to Aaron Bernstein's *Popular Books on Natural Science*. Einstein would write that it was "a work which I read with breathless attention." This book would have a fateful impact on him, because the author included a discussion on the mysteries of electricity. Bernstein asked the reader to take a fanciful ride inside a telegraph wire, racing alongside an electric signal at fantastic speeds.

At the age of sixteen, Einstein had a daydream that led him to an insight which would later change the course of human history. Perhaps remembering the fanciful ride taken in Bernstein's book, Einstein imagined himself running alongside a light beam and asked himself a fateful question: What would the light beam look like? Like Newton visualizing throwing a rock until it orbited the earth like the moon, Einstein's attempt to imagine such a light beam would yield deep and surprising results.

In the Newtonian world, you can catch up to anything if you move fast enough. A speeding car, for example, can race alongside a train. If you peer inside the train, you can see the passengers reading their newspapers and drinking their coffee as if they were sitting in their living rooms. Although they might be hurtling at great speed, they look perfectly stationary as we ride alongside at the same speed in the car.

Similarly, imagine a police car catching up to a speeding motorist. As the police car accelerates and pulls up alongside the car, the police officer can look into the car and wave to the passenger, asking him to pull over. To the officer, the motorist in the car appears at rest, although both the police officer and the motorist may be traveling at a hundred miles an hour.

Physicists knew that light consisted of waves, so Einstein reasoned that if you could run alongside a light beam, then the light beam should be perfectly at rest. This means that the light beam, as seen by the runner, would look like a frozen wave, a still photograph of a wave. It would not oscillate in time. To the young Einstein, however, this did not make any sense. Nowhere had anyone ever seen a frozen wave; there was no such description of one in the scientific literature. Light, to Einstein, was special. You could not catch up to a light beam. Frozen light did not exist.

He did not understand it then, but he accidentally stumbled upon one of the greatest scientific observations of the century, leading up to the principle of relativity. He would later write that "such a principle resulted from a paradox upon which I had already hit at the age of sixteen: If I pursue a beam of light with the velocity c (velocity of light in a vacuum), I should observe such a beam of light . . . at rest. However, there seems to be no such thing, whether on the basis of experience or according to Maxwell's equations."

It was precisely his ability to isolate the key principles behind any phenomena and zero in on the essential picture that put Einstein on the brink of mounting a scientific revolution. Unlike lesser scientists who often got lost in the mathematics, Einstein thought in terms of simple physical pictures—speeding trains, falling elevators, rockets, and moving clocks. These pictures would unerringly guide him through the greatest ideas of the twentieth century. He wrote, "All physical theories, their mathematical expression notwithstanding, ought to lend themselves to so simple a description that even a child could understand."

In the fall of 1895, Einstein finally entered the Polytechnic and began an entirely new phase in his life. For the first time, he thought, he would be exposed to the latest developments in physics that were being debated across the continent. He knew that revolutionary winds were blowing in the world of physics. Scores of new experiments were being performed, seemingly in defiance of the laws of Isaac Newton and classical physics.

At the Polytechnic, Einstein wanted to learn new theories about light, especially Maxwell's equations, which he would later write were the "most fascinating subject at the time I was a student." When Einstein finally learned Maxwell's equations, he could answer the question that was continually on his mind. As he suspected, he found that there were no solutions of Maxwell's equations in which light was frozen in time. But then he discovered more. To his surprise, he found that in Maxwell's theory, light beams always traveled at the same velocity, no matter how fast you moved. Here at last was the final answer to the riddle: *you could never catch up to a light beam because it always sped away from you at the same speed.* This, in turn, violated everything his common sense told him about the world. It would take him several more years to unravel the paradoxes from that key observation, that light always travels at the same velocity.

These revolutionary times demanded revolutionary new theories, and new, daring leaders. Unfortunately, Einstein did not find these leaders at the Polytechnic. His teachers preferred to dwell on classical physics, prompting Einstein to cut his classes and spend most of his time in the laboratory or learning about new theories by himself. His professors viewed these repeated absences from class as chronic laziness; once again Einstein's teachers underestimated him.

Among the teachers at the Polytechnic was physics professor Heinrich Weber, the same person who had been impressed with Einstein and offered to let him audit his classes after he failed the entrance exam. He had even promised Einstein a job as his assistant after graduation. Over time, however, Weber began to resent Einstein's impatience and disregard for authority. Eventually, the professor withdrew his support for Einstein, saying, "You are a smart boy, Einstein, a very smart boy. But you have one great fault: you do not let yourself be told anything." Physics instructor Jean Pernet also was not fond of Einstein. He was insulted when Einstein once threw the lab manual for one of Pernet's classes into the garbage without even looking at it. But Pernet's assistant defended Einstein, stating that although unorthodox, Einstein's solutions were always right. Pernet nevertheless confronted Einstein: "You're enthusiastic, but hopeless at physics. For your own good, you should switch to something else, medicine maybe, literature, or law." Once, because Einstein had torn up the lab instructions, he accidentally caused an explosion that severely injured his right hand, requiring stitches to close the wound. His relations with Pernet had degenerated so badly that Pernet gave Einstein a "1," the lowest possible grade, in his course. Mathematics professor Hermann Minkowski even called Einstein a "lazy dog."

In contrast to his professors' disdain, the friends Einstein made in Zurich would stand loyally by him for the rest of his life. There were only five students in his physics class that year, and he got to know them all. One was Marcel Grossman, a student of mathematics who took careful, meticulous notes of all the lectures. His notes were so good, in fact, that Einstein frequently borrowed them rather than go to class, often getting better scores on the exams than Grossman himself. (Even today, Grossman's notes are preserved at the university.) Grossman confided to Einstein's mother that "something very great" would someday happen to Einstein.

But one person who caught Einstein's attention was another student in his class, Mileva Maric, a woman from Serbia. It was rare to find a physics student from the Balkans, even rarer to find a woman. Mileva was a formidable person who decided by herself to go to Switzerland because it was the only German-speaking country admitting women to the university. She was only the fifth woman to be accepted to study physics at the Polytechnic. Einstein had met his match, a woman who could speak the language of his first love. He found her irresistible and quickly broke off his relationship with Marie Winteler. He daydreamed that he and Mileva would become professors of physics and make great discoveries together. Soon, they were helplessly in love. When they were separated during vacations, they would exchange long, torrid love letters, calling each other by a host of fond nicknames, such as "Johnny" and "Dollie." Einstein would write her poems as well as exhortations of his love: "I can go anywhere I want—but I belong nowhere, and I miss your two little arms and the glowing mouth full of tenderness and kisses." Einstein and Mileva exchanged over 430 letters, preserved by one of their sons. (Ironically, although they lived in near poverty, just one step ahead of the bill collectors, some of these letters recently fetched $400,000 at an auction.)

Einstein's friends could not understand what he saw in Mileva. While Einstein was outgoing with a quick sense of humor, Mileva, four years older, was much darker. She was moody, intensely private, and distrustful of others. She also walked with a noticeable limp due to a congenital problem (one leg was shorter than the other), which further isolated her from others. Friends whispered behind her back about the peculiar behavior of her sister Zorka, who acted strangely and would later become institutionalized as a schizophrenic. But, most important, her social status was questionable. Whereas the Swiss might sometimes look down on Jews, Jews in turn sometimes looked down on southern Europeans, especially from the Balkans.

Mileva, in turn, had no illusions about Einstein. His brilliance was legendary, as well as his irreverent attitude toward authority. She knew that he had renounced his German citizenship and held unpopular views concerning war and peace. She would write, "My sweetheart has a very wicked tongue and is a Jew into the bargain."

Einstein's growing involvement with Mileva, however, was opening up a seismic chasm with his parents. His mother, who had looked approvingly on his relationship with Marie, thoroughly disliked Mileva, regarding her as beneath Albert and someone who would bring ruin to him and their reputation. She was simply too old, too sick, too unfeminine, too gloomy, and too Serbian. "This Miss Maric is causing me the bitterest hours of my life," she confided to a friend. "If it were in my power, I would make every possible effort to banish her from our horizon. I really dislike her. But I have lost all influence with Albert." She would warn him, "By the time you're 30, she'll be an old witch."

But Einstein was determined to see Mileva, even if it meant

causing a deep rupture in his close-knit family. Once, when Einstein's mother was visiting her son, she asked, "What's to become of her?" When Einstein replied, "My wife," she suddenly threw herself on the bed, sobbing uncontrollably. His mother accused him of destroying his future for a woman "who cannot gain entrance to a good family." Eventually, facing the fierce opposition from his parents, Einstein would have to shelve any thoughts of marriage to Mileva until he finished school and got a well-paying job.

In 1900, when Einstein finally graduated from the Polytechnic with a degree in physics and mathematics, his luck soured. It was assumed that he would be given an assistantship. This was the norm, especially since he had passed all his exams and had done well in school. But because Professor Weber had withdrawn his job offer, Einstein was the only one in his class denied an assistantship—a deliberate slap in the face. Once so cocky, he suddenly found himself in uncertain circumstances, especially as financial support from a well-to-do aunt in Genoa dried up with his graduation.

Unaware of the depth of Weber's intense antipathy, Einstein foolishly gave Weber's name as a reference, not realizing that this would further sabotage his future. Reluctantly, he began to realize that this error probably doomed his career even before it started. He would lament bitterly, "I would have found [a job] long ago if Weber had not played a dishonest game with me. All the same, I leave no stone unturned and do not give up my sense of humor. . . . God created the donkey and gave him a thick hide."

Meanwhile, Einstein also applied for Swiss citizenship, but this was impossible until he could prove he was employed. His world was collapsing swiftly. The thought of having to play the violin on the street like a beggar crossed his mind.

His father, realizing his son's desperate plight, wrote a letter to Professor Wilhelm Ostwald of Leipzig, pleading with him to give his son an assistantship. (Ostwald never even responded to this letter. Ironically, a decade later Ostwald would be the first person to nominate Einstein for the Nobel Prize in physics.) Einstein would note how unfair the world suddenly became: "By the mere existence of his stomach, everyone was condemned to participate in that chase." He wrote sadly, "I am nothing but a burden to my relatives. . . . It would surely be better if I did not live at all."

To make matters worse, just at this time his father's business once again went bankrupt. In fact, his father spent all of his wife's inheritance and was deeply in debt to her family. Lacking any financial support, Einstein had no choice but to find the most menial teaching position available. Desperate, he began to comb the newspapers for any hint of a job. At one point, he almost gave up hope of ever becoming a physicist and seriously considered working for an insurance company.

In 1901, he found a job teaching mathematics at the Winterthur Technical School. Somehow, between exhausting teaching duties, he was able to squeeze enough time to write his first published paper, "Deductions from the Phenomena of Capillarity," which Einstein himself realized was not earthshaking. The next year, he took a temporary tutoring position at a boarding school in Schaffhausen. True to form, he could not get along with the authoritarian director of the school, Jakob Nuesch, and was summarily fired. (The director was so inflamed that he even accused Einstein of fomenting a revolution.)

Einstein was beginning to think that he would spend the rest of his life forced to eke out a menial existence tutoring indifferent students and scouring the ads in newspapers. His friend Friedrich Adler would recall that at this time Albert was close to starvation. He was a complete failure. Still he refused to ask his relatives for a handout. Then he faced two more setbacks. First, Mileva flunked her final exams at the Polytechnic for the second time. This meant that her career as a physicist was essentially finished. No one would ever accept her into a graduate program with her dismal record. Painfully disheartened, she lost interest in physics. Their romantic dreams of exploring the universe together were over. And then, in November 1901, when Mileva was back home, he received a letter from her telling him that she was pregnant!

Einstein, despite his lack of prospects, was still thrilled at the possibility of becoming a father. Being separated from Mileva was torture, but they furiously exchanged letters, almost daily. On February 4, 1902, he finally learned he was a father of a baby girl, born at Mileva's parents' home in Novi Sad and christened Lieserl. Excited, Einstein wanted to know everything about her. He even pleaded with Mileva to please send a photo or a sketch of her. Mysteriously, no one is sure what happened to the child. The last mention of her is in a letter from September 1903, which stated that she was suffering from scarlet fever. Historians believe that she most likely died of the disease or that she was eventually given up for adoption.

Just when his fortunes appeared as if they could not sink any lower, Einstein received word from an unexpected source. His good friend Marcel Grossman had been able to secure a job for him as a minor civil servant at the Bern Patent Office. From that lowly position, Einstein would change the world. (To keep alive his fading hopes of one day becoming a professor, he persuaded Professor Alfred Kleiner of the University of Zurich to be his Ph.D. advisor during this period.)

On June 23, 1902, Einstein began work at the patent office as a technical expert, third class, with a paltry salary. In hindsight, there were at least three hidden advantages to working at this office. First, his job forced him to find the basic physical principles that underlay any invention. During the day, he honed his considerable physical instincts by stripping away unnecessary details and isolating the essential ingredient of each patent and then writing up a report. His reports were so long on detail and analysis that he wrote to his friends that he was making a living "pissing ink." Second, many of the patent applications concerned electromechanical devices, so his ample experience visualizing the inner workings of dynamos and electric motors from his father's factory was a great help. And last, it freed him from distractions and gave him time to ponder deep questions about light and motion. Often, he could finish the details of his work quickly, leaving hours to fall back on the daydreams that dogged him since his youth. In his work and especially at night, he returned to his physics. The quiet atmosphere of the patent office suited him. He called it his "worldly monastery."

Einstein had barely settled into his new job at the patent office when he learned that his father was dying of heart disease. In October he had to depart immediately for Milan. On his deathbed, Hermann finally gave Albert his blessing to marry Mileva. The death left Albert with an overpowering sense that he had disappointed his father and family, a feeling that would stay with him forever. His secretary, Helen Dukas, wrote, "Many years later, he still recalled vividly his shattering sense of loss. Indeed on one occasion he wrote that his father's death was the deepest shock he had ever experienced." Maja, in particular, bitterly noted that "sad fate did not permit [her father] even to suspect that two years later his son would lay the foundation of his future greatness and fame."

In January of 1903, Einstein finally felt secure enough to marry Mileva. A year later, their son Hans was born. Einstein settled down to the life of a lowly civil servant in Bern, a husband, and a father. His friend David Reichinstein vividly remembered visiting Einstein during this period: "The door of the flat was open to allow the floor, which had just been scrubbed, and the washing, hung up in the hall, to dry. I entered Einstein's room. With one hand, he was stoically rocking a bassinet in which there was a child. In his mouth, Einstein had a bad, a very bad cigar, and in the other hand, an open book. The stove was smoking horribly."

To raise some extra money, he put an ad in the local newspaper, offering "private lessons in mathematics and physics." It was the first recorded mention of Einstein's name in any newspaper. Maurice Solovine, a Jewish Romanian student of philosophy, was the first student to answer the ad. To his delight, Einstein found that Solovine was an excellent sounding board for his many ideas about space, time, and light. To prevent himself from becoming dangerously isolated from the main currents in physics, he hit upon the idea of forming an informal study group, which Einstein mockingly called the "Olympian Academy," to debate the great issues of the day.

In hindsight, the days spent with the academy group were perhaps the most joyous in Einstein's life. Decades later, tears would come to his eyes when he recalled the vibrant, audacious claims they made as they voraciously devoured all the major scientific works of the day. Their spirited debates and raucous discussions filled the coffeehouses and beer halls of Zurich, and anything seemed possible. They would fondly swear, "These words of Epicurus applied to us: 'What a beautiful thing joyous poverty is!' "

In particular, they pored over the controversial work of Ernst Mach, a Viennese physicist and philosopher who was something of a gadfly, challenging any physicist who spoke of things that were beyond our senses. Mach spelled out his theories in an influential book, *The Science of Mechanics*. He challenged the idea of atoms, which he thought were hopelessly beyond the realm of measurement. What most riveted Einstein's attention, however, was Mach's scathing criticism of the aether and absolute motion. To Mach, the imposing edifice of Newtonian mechanics was based on sand, as the concepts of absolute space and absolute time could never be measured. He believed relative motions could be measured, but absolute motions could not. No one had ever found the mystical absolute reference frame that could determine the motion of the planets and the stars, and no one had ever found even the slightest experimental evidence for the aether either.

One series of experiments that indicated a fatal weakness in this Newtonian picture had been performed in 1887 by Albert Michelson and Edward Morley, who had set out to give the best possible measurement of the properties of this invisible aether. They reasoned that the earth moves within this sea of aether, creating an "aether wind," and hence the speed of light should change, depending on the direction the earth took.

Think, for example, of running with the wind. If you run in the same direction as the wind, then you feel yourself being pushed along with the wind. With the wind at your back, you travel at a faster speed, and in fact your speed has been increased by the speed of the wind. If you run into the wind, then you slow down; your speed is now decreased by the speed of the wind. Similarly if you run sideways, 90 degrees to the wind, you are blown off to the side with yet another speed. The point is that your speed changes depending on which direction you run with respect to the wind.

Michelson and Morley devised a clever experiment whereby a single beam of light is split into two distinct beams, each shot in different directions at right angles to each other. Mirrors reflected the beams back to the source, and then the two beams were allowed to mix and interfere with each other. The whole apparatus was carefully placed on a bed of liquid mercury, so that it could rotate freely, and it was so delicate that it easily picked up the motion of passing horse carriages. According to the aether theory, the two beams should travel at different speeds. One beam, for example, would move along the direction of the earth's motion in the aether, and the other beam would move at 90 degrees to the aether wind. Thus, when they returned back to the source, they should be out of phase with each other.

Much to their astonishment, Michelson and Morley found

that the speed of light was identical for all light beams, no matter in which direction the apparatus pointed. This was deeply disturbing, for it meant that there was no aether wind at all, and the speed of light never changed, even as they rotated their apparatus in all directions.

This left physicists with two equally unpleasant choices. One was that the earth might be perfectly stationary with respect to the aether. This choice seemed to violate everything known about astronomy since the original work of Copernicus, who found that there was nothing special about the location of the earth in the universe. Second, one might abandon the aether theory and Newtonian mechanics along with it.

Heroic efforts were made to salvage the aether theory. The closest step toward a resolution to this puzzle was found by the Dutch physicist Hendrik Lorentz and the Irish physicist George FitzGerald. They reasoned that the earth, in its motion through the aether, was actually physically compressed by the aether wind, so that all meter sticks in the Michelson-Morley experiment were shrunken. The aether, which already had near mystical properties of being invisible, noncompressible, extremely dense, and so on, now had one more property: it could mechanically compress atoms by passing through them. This would conveniently explain the negative result. In this picture, the speed of light did in fact change, but you could never measure it because every time you tried using a meter stick, the velocity of light would indeed change and the meter sticks would shrink in the direction of the aether wind by precisely the right amount.

Lorentz and FitzGerald independently calculated the amount of shrinkage, yielding what is now called the "Lorentz-FitzGerald contraction." Neither Lorentz nor FitzGerald were especially pleased with this result; it was simply a quick fix, a way of patching up Newtonian mechanics, but it was the best they could do. Not many physicists liked the Lorentz-FitzGerald contraction either, since it smacked of being an ad hoc principle, thrown in to prop up the sagging fortunes of the aether theory. To Einstein, the idea of the aether, with its near miraculous properties, seemed artificial and contrived. Earlier, Copernicus had destroyed the earth-centered solar system of Ptolemy, which required the planets to move in extremely complex circular motions called "epicycles." With Occam's Razor, Copernicus sliced away the blizzard of epicycles needed to patch up the Ptolemaic system and put the sun at the center of the solar system.

Like Copernicus, Einstein would use Occam's Razor to slice away the many pretensions of the aether theory. And he would do this by using a children's picture.

CHAPTER 3

Special Relativity and the "Miracle Year"

Intrigued by Mach's criticisms of Newton's theory, Einstein went back to the picture that had haunted him since he was sixteen, running alongside a light beam. He returned to the curious but important discovery that he made while at the Polytechnic, that in Maxwell's theory the speed of light was the same, no matter how you measured it. For years, he puzzled over how this could possibly happen, because in a Newtonian, commonsense world you can always catch up to a speeding object.

Imagine again the police officer chasing after a speeding motorist. If he drives fast enough, the officer knows that he can catch up to the motorist. Anyone who has ever gotten a ticket for speeding knows that. But if we now replace the speeding motorist with a light beam, and an observer witnesses the whole thing, then the observer concludes that the officer is speeding just behind the light beam, traveling almost as fast as light. We are confident that the officer knows he is traveling neck and neck with the light beam. But later, when we interview him, we hear a strange tale. He claims that instead of riding alongside the light beam as we just witnessed, it sped away from him, leaving him in the dust. He says that no matter how much he gunned his engines, the light beam sped away at precisely the same velocity. In fact, he swears that he could not even make a dent in catching up to the light beam. No matter how fast he traveled, the light beam traveled away from him at the speed of light, as if he were stationary instead of speeding in a police car.

But when you insist that you saw the police officer speeding neck and neck with the light beam, within a hairsbreadth of catching up to it, he says you are crazy; he never even got close. To Einstein, this was the central, nagging mystery: *How was it possible for two people to see the same event in such totally different ways?* If the speed of light was really a constant of nature, then how could a witness claim that the officer was neck and neck with the light beam, yet the officer swears that he never even got close?

Einstein had realized earlier that the Newtonian picture (where velocities can be added and subtracted) and the Maxwellian picture (where the speed of light was a constant) were in total contradiction. Newtonian theory was a self-contained system, resting on a few assumptions. If only one of these assumptions were changed, it would unravel the entire theory in the same way that a loose thread can unravel a sweater. That thread would be Einstein's daydream of racing a light beam.

One day around May of 1905, Einstein went to visit his good friend Michele Besso, who also worked at the patent office, and laid out the dimensions of the problem that had puzzled him for a decade. Using Besso as his favorite sounding board for ideas, Einstein presented the issue: Newtonian mechanics and Maxwell's equations, the two pillars of physics, were incompatible. One or the other was wrong. Whichever theory proved to be correct, the final resolution would require a vast reorganization of all of physics. He went over and over the paradox of racing a light beam. Einstein would later recall, "The germ of the special relativity theory was already present in that paradox." They talked for hours, discussing every aspect of the problem, including Newton's concept of absolute space and time, which seemed to violate Maxwell's constancy of the speed of light.

Eventually, totally exhausted, Einstein announced that he was defeated and would give up the entire quest. It was no use; he had failed.

Although Einstein was depressed, his thoughts were still churning in his mind when he returned home that night. In particular, he remembered riding in a street car in Bern and looking back at the famous clock tower that dominated the city. He then imagined what would happen if his street car raced away from the clock tower at the speed of light. He quickly realized that the clock would appear stopped, since light could not catch up to the street car, but his own clock in the street car would beat normally.

Then it suddenly hit him, the key to the entire problem. Einstein recalled, "A storm broke loose in my mind." The answer was simple and elegant: *time can beat at different rates throughout the universe, depending on how fast you moved.* Imagine clocks scattered at different points in space, each one announcing a different time, each one ticking at a different rate. One second on Earth was not the same length as one second on the moon or one second on Jupiter. In fact, the faster you moved, the more time slowed down. (Einstein once joked that in relativity theory, he placed a clock at every point in the universe, each one running at a different rate, but in real life he didn't have enough money to buy even one.) This meant that events that were simultaneous in one frame were not necessarily simultaneous in another frame, as Newton thought. He had finally tapped into "God's thoughts." He would recall excitedly, "The solution came to me suddenly with the thought that our concepts and laws of space and time can only claim validity insofar as they stand in a clear relation to our experiences. . . . By a revision of the concept of simultaneity into a more malleable form, I thus arrived at the theory of relativity."

For example, remember that in the paradox of the speeding motorist, the police officer was traveling neck and neck with the speeding light beam, while the officer himself claimed that the light beam was speeding away from him at precisely the speed of light, no matter how much he gunned his engines. The only way to reconcile these two pictures is to have the brain of the officer slow down. *Time slows down for the policeman*. If we could have seen the officer's wristwatch from the roadside, we would have seen that it nearly stopped and that his facial expressions were frozen in time. Thus, from our point of view, we saw him speeding neck and neck with the light beam, but his clocks (and his brain) were nearly stopped. When we interviewed the officer later, we found that he perceived the light beam to be speeding away, only because his brain and clocks were running much slower.

To complete his theory, Einstein also incorporated the Lorentz-FitzGerald contraction, except that it was space itself that was contracted, not the atoms, as Lorentz and FitzGerald thought. (The combined effect of space contraction and time dilation is today called the "Lorentz transformation.") Thus, he could dispense entirely with the aether theory. Summarizing the path that he took to relativity, he would write, "I owe more to Maxwell than to anyone." Apparently, although he was dimly aware of the Michelson-Morley experiment, the inspiration for relativity did not come from the aether wind, but directly from Maxwell's equations.

The day after this revelation, Einstein went back to Besso's home and, without even saying hello, he blurted out, "Thank you, I've completely solved the problem." He would proudly recall, "An analysis of the concept of time was my solution. Time cannot be absolutely defined, and there is an inseparable relation between time and signal velocity." For the next six weeks, he furiously worked out every mathematical detail of his brilliant insight, leading to a paper that is arguably one of the most important scientific papers of all time. According to his son, he then went straight to bed for two weeks after giving the paper to

Mileva to check for any mathematical errors. The final paper, "On the Electrodynamics of Moving Bodies," was scribbled on thirty-one handwritten pages, but it changed world history.

In the paper, he does not acknowledge any other physicist; he only gives thanks to Michele Besso. (Einstein was aware of Lorentz's early work on the subject, but not the Lorentz contraction itself, which Einstein found independently.) It was finally published in *Annalen der Physik* in September 1905, in volume 17. In fact, Einstein would publish three of his pathbreaking papers in that famous volume 17. His colleague Max Born has written, volume 17 is "one of the most remarkable volumes in the whole scientific literature. It contains three papers by Einstein, each dealing with a different subject and each today acknowledged to be a masterpiece." (Copies of that famous volume recently sold for $15,000 at an auction in 1994.)

With almost breathtaking sweep, Einstein began his paper by proclaiming that his theories worked not just for light, but were truths about the universe itself. Remarkably, he derived all his work from two simple postulates applying to inertial frames (i.e., objects that move with constant velocity with respect to each other):

1. *The laws of physics are the same in all inertial frames.*
2. *The speed of light is a constant in all inertial frames.*

These two deceptively simple principles mark the most profound insights into the nature of the universe since Newton's work. From them, one can derive an entirely new picture of space and time.

First, in one masterful stroke, Einstein elegantly proved that if the speed of light was indeed a constant of nature, then the most general solution was the Lorentz transformation. He then showed that Maxwell's equations did indeed respect that principle. Last, he showed that velocities add in a peculiar way.

Although Newton, observing the motion of sailing ships, concluded that velocities could add without limit, Einstein concluded that the speed of light was the ultimate velocity in the universe. Imagine, for the moment, that you are in a rocket speeding at 90% the speed of light away from Earth. Now fire a bullet inside the rocket that is also going at 90% the speed of light. According to Newtonian physics, the bullet should be going at 180% the speed of light, thus exceeding light velocity. But Einstein showed that because meter sticks are shortening and time is slowing down, the sum of these velocities is actually close to 99% the speed of light. In fact, Einstein could show that no matter how hard you tried, you could never boost yourself beyond the speed of light. Light velocity was the ultimate speed limit in the universe.

We never see these bizarre distortions in our experience because we never travel near the speed of light. For everyday velocities, Newton's laws are perfectly fine. This is the fundamental reason why it took over two hundred years to discover the first correction to Newton's laws. But now imagine that the speed of light is only 20 miles per hour. If a car were to go down the street, it might look compressed in the direction of motion, being squeezed like an accordion down to perhaps 1 inch in length, for example, although its height would remain the same. Because the passengers in the car are compressed down to 1 inch, we might expect them to yell and scream as their bones are crushed. In fact, the passengers see nothing wrong, since everything inside the car, including the atoms in their bodies, is squeezed as well.

As the car slows down to a stop, it would slowly expand from 1 inch to about 10 feet, and the passengers would walk out as if nothing happened. Who is really compressed? You or the car? According to relativity, you cannot tell, since the concept of length has no absolute meaning.

In retrospect, one can see that others came tantalizingly close to discovering relativity. Lorentz and FitzGerald obtained the same contraction, but they had the totally wrong understanding of the result, thinking it was an electromechanical deformation of the atoms, rather than a subtle transformation of space and time itself. Henri Poincaré, recognized as the greatest French mathematician of his time, came close. He understood that the speed of light must be a constant in all inertial frames, and even showed that Maxwell's equations retained the same form under a Lorentz transformation. However, he too refused to abandon the Newtonian framework of the aether and thought that these distortions were strictly a phenomenon of electricity and magnetism.

Einstein then pushed further and made the next fateful leap. He wrote a small paper, almost a footnote, late in 1905 that would change world history. If meter sticks and clocks became distorted the faster you moved, then everything you can measure with meter sticks and clocks must also change, including matter and energy. In fact, matter and energy could change into each other. For example, Einstein could show that the mass of an object increased the faster it moved. (Its mass would in fact become infinite if you hit the speed of light—which is impossible, which proves the unattainability of the speed of light.) This meant that the energy of motion was somehow being transformed into increasing the mass of the object. *Thus, matter and energy are interchangeable.* If you calculated precisely how much energy was being converted into mass, in a few simple lines you could show that $E=mc^2$, the most celebrated equation of all time. Since the speed of light was a fantastically large number, and its square was even larger, this meant that even a tiny amount of matter could release a fabulous amount of energy. A few teaspoons of matter, for example, has the energy of several hydrogen bombs. In fact, a piece of matter the size of a house might be enough to crack the earth in half.

Einstein's formula was not simply an academic exercise,

because he believed that it might explain a curious fact discovered by Marie Curie, that just an ounce of radium emitted 4,000 calories of heat per hour indefinitely, seemingly violating the first law of thermodynamics (which states that the total amount of energy is always constant or conserved). He concluded that there should be a slight decrease in its mass as radium radiated away energy (an amount too small to be measured using the equipment of 1905). "The idea is amusing and enticing; but whether the Almighty is laughing at it and is leading me up the garden path—that I cannot know," he wrote. He concluded that a direct verification of his conjecture "for the time being probably lies beyond the realm of possible experience."

Why hadn't this untapped energy been noticed before? He compared this to a fabulously rich man who kept his wealth secret by never spending a cent.

Banesh Hoffman, a former student, wrote, "Imagine the audacity of such a step. . . . Every clod of earth, every feather, every speck of dust becoming a prodigious reservoir of untapped energy. There was no way of verifying this at the time. Yet in presenting his equation in 1907 Einstein spoke of it as the most important consequence of his theory of relativity. His extraordinary ability to see far ahead is shown by the fact that his equation was not verified . . . until some twenty-five years later."

Once again, the relativity principle forced a major revision in classical physics. Before, physicists believed in the conservation of energy, the first law of thermodynamics, which states that the total amount of energy can never be created or destroyed. Now, physicists considered the total combined amount of matter and energy as being conserved.

Einstein's restless mind tackled yet another problem that same year, the photoelectric effect. Heinrich Hertz, back in 1887, noticed that if a light beam struck a metal, under certain circumstances a small electric current could be created. This is the same principle that underlies much of modern electronics. Solar cells convert ordinary sunlight into electrical power, which can be used to energize our calculators. TV cameras take light beams from the subject and convert them into electric currents, which eventually wind up on our TV screen.

However, at the turn of the century, this was still a total mystery. Somehow, the light beam was knocking electrons out of the metal, but how? Newton had believed that light consisted of tiny particles that he called "corpuscles," but physicists were convinced that light was a wave, and according to classical wave theory its energy was independent of its frequency. For example, although red and green light have different frequencies, they should have the same energy, and hence, when they hit a piece of metal, the energy of the ejected electrons should be the same as well. Similarly, classical wave theory said that if one turned up the intensity of the light beam by adding more lamps, then the energy of the ejected electrons should increase. The work of Philipp Lenard, however, demonstrated that the energy of the ejected electrons was strictly dependent on the frequency or color of the light beam, not its intensity, which was opposite the prediction of the wave theory.

Einstein sought to explain the photoelectric effect by using the new "quantum theory" recently discovered by Max Planck in Berlin in 1900. Planck made one of the most radical departures from classical physics by assuming that energy was not a smooth quantity, like a liquid, but occurred in definite, discrete packets, called "quanta." The energy of each quantum was proportional to its frequency. The proportionality constant was a new constant of nature, now called "Planck's constant." One reason why the world of the atom and the quantum seems so bizarre is the fact that Planck's constant is a very small number. Einstein reasoned that if energy occurred in discrete packets, then light itself must be quantized. (Einstein's packet of "light quanta" was later christened the "photon," a particle of light, by chemist Gilbert Lewis in 1926.) Einstein reasoned that if the energy of the photon was proportional to its frequency, then the energy of the ejected electron should also be proportional to its frequency, contrary to classical physics. (It's amusing to note that on the popular TV series *Star Trek*, the crew of the *Enterprise* fires "photon torpedoes" at its enemies. In reality, the simplest launcher of photon torpedoes is a flashlight.)

Einstein's new picture, a quantum theory of light, made a direct prediction that could be tested experimentally. By turning up the frequency of the incoming light beam, one should be able to measure a smooth rise in the voltage generated in the metal. This historic paper (which would eventually win him the Nobel Prize in physics) was published on June 9, 1905, with the title "On a Heuristic Point of View Concerning the Production and Transformation of Light." With it, the photon was born, as well as the quantum theory of light.

In yet another article written during the "miracle year" of 1905, Einstein tackled the problem of the atom. Although the atomic theory was remarkably successful in determining the properties of gases and chemical reactions, there was no direct proof of their existence, as Mach and other critics were fond of pointing out. Einstein reasoned that one might be able to prove the existence of atoms by noticing their effect on small particles in a liquid. "Brownian motion," for example, refers to tiny, random motions of small particles suspended in a liquid. This property was discovered in 1828 by the botanist Robert Brown, who observed tiny pollen grains under a microscope exhibiting strange random motions. At first, he thought that these zigzag movements were like the motion of male sperm cells. But he found that this strange aberrant behavior was also exhibited by tiny grains of glass and granite.

Some had speculated that Brownian motion might be due to the random impacts of molecules, but no one could formulate a reasonable theory of this. Einstein, however, took the next decisive step. He reasoned that although atoms were too small to be observed, one could estimate their size and behavior by calculating their cumulative impact on large objects. If one seriously believed in the theory of atoms, then the atomic theory should be able to calculate the physical dimensions of atoms by analyzing Brownian motion. By assuming that the random collisions of trillions upon trillions of molecules of water were causing random motions of a dust particle, he was able to compute the size and weight of atoms, thereby giving experimental evidence of the existence of atoms.

It was nothing short of amazing that by peering into a simple microscope, Einstein could calculate that a gram of hydrogen contained 3.03×10^{23} atoms, which is close to the correct value. The title was "On the Movement of Small Particles Suspended in Stationary Liquids Required by the Molecular-Kinetic Theory of Heat" (July 18). This simple paper, in effect, gave the first experimental proof of the existence of atoms. (Ironically, just a year after Einstein calculated the size of atoms, the physicist Ludwig Boltzmann, who had pioneered the theory of atoms, committed suicide, in part because of the incessant ridicule he received for advancing the atomic theory.) After Einstein wrote these four historic papers, he also submitted an earlier paper of his, on the size of molecules, to his advisor, Professor Alfred Kleiner, as his dissertation. That night, he got drunk with Mileva.

At first, his dissertation was refused. But on January 15, 1906, Einstein was finally granted a Ph.D. from the University of Zurich. He could now call himself Dr. Einstein. The birth of the new physics all took place at the Einstein residence on Kramgasse 49 in Bern. (Today, it is called the "Einstein House." As you gaze through its beautiful bay window facing the street, you can read a plaque which states that through this window, the theory of relativity was created. On the other wall, you can see a picture of the atomic bomb.)

Thus, 1905 was truly an *annus mirabilis* in the history of science. If we are to find another comparable miracle year, we would have to look back to 1666, when Isaac Newton, twenty-three years old, hit upon the universal law of gravitation, the integral and differential calculus, the binomial theorem, and his theory of color.

Einstein had finished the year 1905 laying down the photon theory, providing evidence for the existence of atoms, and toppling the framework of Newtonian physics, each of them worthy of international acclaim. He was disappointed, however, in the deafening silence that ensued. His work, it seemed, was totally ignored. Disheartened, Einstein went about his personal life, raising his child and toiling by himself at the patent office. Perhaps the thought of pioneering new worlds in physics was all a pipe dream.

In early 1906, however, the first glimmer of response riveted Einstein's attention. He got just a single letter, but it was from perhaps the most important physicist of the time, Max Planck, who immediately understood the radical implications of Einstein's work. What drew Planck to the theory of relativity was that it elevated a quantity, the speed of light, into a fundamental constant of nature. Planck's constant, for example, demarcated the classical world from the subatomic world of the quantum. We are shielded from the strange properties of atoms because of the smallness of Planck's constant. Similarly, felt Planck, Einstein was raising the speed of light into a new constant of nature. We were shielded from the equally bizarre world of cosmic physics by the huge value of the speed of light.

In Planck's mind, these two constants, Planck's constant and the speed of light, laid down the limits of common sense and Newtonian physics. We cannot see the fundamentally weird nature of physical reality because of the smallness of Planck's constant and the immensity of the speed of light. If relativity and the quantum theory violated common sense, it was only because we live our entire life in a tiny corner of the universe, in a sheltered world where velocities are low compared to the speed of light and objects are so large we never encounter Planck's constant. Nature, however, does not care about our common sense, but created a universe based on subatomic particles that routinely go near the speed of light and obey Planck's formula.

In the summer of 1906, Planck sent his assistant, Max von Laue, to meet this obscure civil servant who appeared out of nowhere to challenge the legacy of Isaac Newton. They were supposed to meet in the waiting room of the patent office but comically walked right past each other because von Laue expected to see an imposing, authoritative figure. When Einstein finally introduced himself, von Laue was surprised to find someone completely different, a surprisingly young and casually dressed civil servant. They became lifelong friends. (However, von Laue knew a bad cigar when he saw one. When Einstein offered him a cigar, von Laue discreetly threw it into the Aare River when Einstein wasn't looking as they talked and crossed a bridge.)

With the blessing of Max Planck, the work of Einstein gradually began to attract the attention of other physicists. Ironically, one of Einstein's old professors at the Polytechnic, who had called him a "lazy dog" for cutting his classes, took a particular interest in the work of his former student. The mathematician Hermann Minkowski plunged ahead and developed the equations of relativity even further, trying to reformulate Einstein's observation that space turns into time and vice versa the faster you move. Minkowski put this into mathematical language and concluded that space and time formed a four-dimensional unity. Suddenly, everyone was talking about the fourth dimension.

For example, on a map, two coordinates (length and width) are required to locate any point. If you add a third dimension, height, then you can locate any object in space, from the tip of your nose to the ends of the universe. The visible world around us is thus three-dimensional. Writers like H. G. Wells had conjectured that perhaps time could be viewed as a fourth dimension, such that any event could be located by giving its three-dimensional coordinates and the time at which it occurs. Thus, if you want to meet someone in New York City, you might say, "Meet me at the corner of 42nd Street and Fifth Avenue, on the twentieth floor, at noon." Four numbers uniquely specify the event. But Wells's fourth dimension was only an idea without any mathematical or physical content.

Minkowski then rewrote Einstein's equations to reveal this beautiful four-dimensional structure, forever linking space and time into a four-dimensional fabric. Minkowski wrote, "From now on, space and time separately have vanished into the merest shadows, and only a kind of union of the two will preserve any independent reality."

At first, Einstein was not impressed. In fact, he even wrote derisively, "The main thing is the content, not the mathematics. With mathematics, you can prove anything." Einstein believed that at the core of relativity lay basic physical principles, not pretty but meaningless four-dimensional mathematics, which he called "superfluous erudition." To him, the essential thing was to have a clear and simple picture (e.g., trains, falling elevators, rockets), and the mathematics would come later. In fact, at this point he thought that mathematics represented only the bookkeeping necessary to track what was happening in the picture.

Einstein would write, half in jest, "Since the mathematicians have attacked the relativity theory, I myself no longer understand it anymore." With time, however, he began to appreciate the full power of Minkowski's work and its deep philosophical implications. What Minkowski had shown was that it was possible to unify two seemingly different concepts by using the power of symmetry. Space and time were now to be viewed as different states of the same object. Similarly, energy and matter, as well as electricity and magnetism, could be related via the fourth dimension. *Unification through symmetry* became one of Einstein's guiding principles for the rest of his life.

Think of a snowflake, for example. If you rotate the snowflake by 60 degrees, the snowflake remains the same. Mathematically, we say that objects that maintain their form after a rotation are called "covariant." Minkowski showed that Einstein's equations, like a snowflake, remain covariant when space and time are rotated as four-dimensional objects.

In other words, a new principle of physics was being born, and it further refined the work of Einstein: *The equations of physics must be Lorentz covariant* (i.e., maintain the same form under a Lorentz transformation). Einstein would later admit that without the four-dimensional mathematics of Minkowski, relativity "might have remained stuck in its diapers." What was remarkable was that this new four-dimensional physics allowed physicists to compress all the equations of relativity into a remarkably compact form. For example, every electrical engineering student and physicist, when first studying Maxwell's series of eight partial differential equations, finds that they are fiendishly difficult. But Minkowski's new mathematics collapsed Maxwell's equations down to just two. (In fact, one can prove using four-dimensional mathematics that Maxwell's equations are the *simplest* possible equations describing light.) For the first time, physicists appreciated the power of symmetry in their equations. When a physicist talks about "beauty and elegance" in physics, what he or she often really means is that symmetry allows one to unify a large number of diverse phenomena and concepts into a remarkably compact form. *The more beautiful an equation is, the more symmetry it possesses, and the more phenomena it can explain in the shortest amount of space.*

Thus, the power of symmetry allows us to unify disparate pieces into their harmonious, integral whole. Rotations of a snowflake, for example, allow us to see the unity that exists between each point on the snowflake. Rotating in four-dimensional space unifies the concept of space and time, turning one into the other as the velocity is increased. This beautiful, elegant concept, that symmetry unifies seemingly dissimilar entities into a pleasing, harmonious whole, guided Einstein for the next fifty years.

Paradoxically, as soon as Einstein completed the theory of special relativity, he began to lose interest, preferring to contemplate another, deeper question, the problem of gravity and acceleration, which seemed beyond the reach of special relativity. Einstein had given birth to relativity theory, but like any loving parent, he immediately realized its potential faults and tried to correct them. (More will be said about this later.)

Meanwhile, experimental evidence began to confirm some of his ideas, raising his visibility within the physics community. The Michelson-Morley experiment was repeated, each time yielding the same negative result and casting doubt on the entire aether theory. Meanwhile, experiments on the photoelectric effect confirmed Einstein's equations. Furthermore, in 1908 experiments on high-speed electrons seemed to prove that the mass of the electron increased the faster it moved. Buoyed by the experimental successes slowly piling up for his theories, he applied for a lectureship (*privatdozent*) position at the nearby University of Bern. This position was below that of a professor, but it offered the advantage that he could simultaneously continue with his job at the patent office. He submitted his relativity thesis as well as other published works. At first, he was rejected by the department head, Aime Foster, who declared that relativity theory was incomprehensible. His second try was successful.

In 1908, with evidence mounting that Einstein had made major breakthroughs in physics, he was seriously considered for a more prestigious position at the University of Zurich. He faced stiff competition, however, from an old acquaintance, Friedrich

Adler. Both top candidates for the position were Jewish, which was a handicap, but Adler was the son of the founder of the Austrian Social Democratic Party, to which many faculty members were sympathetic, so it appeared as if Einstein would be passed over for the position. It was surprising, therefore, that Adler himself argued strongly for Einstein to take the position. He was a shrewd observer of character and sized up Einstein correctly. He wrote eloquently of Einstein's outstanding abilities as a physicist, but noted, "As a student he was treated contemptuously by the professors. . . . He has no understanding of how to get on with important people." Due to Adler's extraordinary self-sacrifice, Einstein was awarded the position and began his meteoric climb up the academic ladder. He now returned to Zurich, but this time as a professor, not as a failed, unemployed physicist and misfit. When he found an apartment in Zurich, he was delighted to learn that Adler lived one floor just below his, and they became good friends.

In 1909, Einstein gave his maiden lecture at his first major physics conference in Salzburg, where many luminaries, including Max Planck, were in attendance. In his talk, "The Development of Our Views on the Nature and Constitution of Radiation," he forcefully brought the equation $E=mc^2$ to the world. Einstein, used to scrimping on funds for his lunch, marveled at the opulence of this conference. He recalled, "The festivities ended in the Hotel National, with the most opulent banquet I have ever attended in my life. It encouraged me to say to the Genevan patrician sitting next to me: Do you know what Calvin would have done had he been here? . . . He would have erected an enormous stake and had us burnt for sinful extravagance. The man never addressed another word to me."

Einstein's talk was the first time in history that anyone had clearly presented the concept of "duality" in physics, the concept that light can have dual properties, either as a wave, as Maxwell had suggested in the previous century, or as a particle, as Newton had suggested. Whether one saw light as a particle or as a wave depended on the experiment. For low-energy experiments, where the wavelength of the light beam is large, the wave picture was more useful. For high-energy beams, where the wavelength of light is extremely small, the particle picture was more suitable. This concept (which decades later would be attributed to Danish physicist Niels Bohr) proved to be a fundamental observation of the nature of matter and energy and one of the richest sources of research on the quantum theory.

Although now a professor, Einstein remained just as bohemian as ever. One student vividly recalled his maiden lecture at the University of Zurich: "He appeared in class in somewhat shabby attire, wearing pants that were too short and carrying with him a slip of paper the size of a visiting card on which he had sketched his lecture notes."

In 1910, Einstein's second son, Eduard, was born. Einstein, ever the restless wanderer, was already looking for a new position, apparently because some professors wanted to remove him from the university. The next year, he was offered a position at the German University of Prague's Institute of Theoretical Physics at an increased salary. Ironically, his office was next to an insane asylum. Pondering the mysteries of physics, he often wondered if the inmates were the sane ones.

That same year, 1911, also marked the meeting of the First Solvay Conference in Brussels, funded by a wealthy Belgian industrialist, Ernest Solvay, which would highlight the work of the world's leading scientists. This was the most important conference of its time, and it gave Einstein a chance to meet the giants of physics and exchange ideas with them. He met Marie Curie, the two-time Nobel laureate, and formed a lifelong relationship. The theory of relativity and his photon theory were the center of attention. The theme of the conference was "The Theory of Radiation and the Quanta."

One question debated at the conference was the famous "twin

paradox." Einstein had already made mention of the strange paradoxes concerning the slowing down of time. The twin paradox was proposed by physicist Paul Langevin, who announced a simple thought experiment that probed some of the seeming contradictions in relativity theory. (At the time, the newspapers were filled with lurid stories about Langevin, who was unhappily married and involved in a scandalous romance with Marie Curie, a widow.) Langevin considered two twins living on Earth. One twin is transported near the speed of light and then returns back to Earth. Fifty years, say, may have transpired on Earth, but since time slows down on the rocket, the rocket twin has aged by only ten years. When the twins finally meet, there is a mismatch in their ages, with the rocket twin being forty years younger.

Now view the situation from the point of view of the rocket twin. From his perspective, he is at rest, and it is Earth that has blasted off, so the earth twin's clocks become slower. When the two twins finally meet, the earth twin should be younger, not the rocket twin. But since motions should be relative, the question is, which twin is really younger? Since the two situations seem symmetrical, this puzzle even today remains a thorn in the side of any student who has tried to tackle relativity.

The resolution of the puzzle, as Einstein pointed out, is that the rocket twin, not the earth twin, has accelerated. The rocket has to slow down, stop, and then reverse, which clearly causes great stress on the rocket twin. In other words, the situations are not symmetrical because accelerations, which are not covered by the assumptions behind special relativity, only occur for the rocket twin, who is really younger.

(However, the situation becomes more puzzling if the rocket twin never returns. In this scenario, each sees by telescope the other twin slowing down in time. Since the situations are now perfectly symmetrical, then each twin is convinced that the other one is younger. Likewise, each twin is convinced that the other is compressed. So which twin is younger and thinner? As paradoxical as it seems, in relativity theory it is possible to have two twins, each younger than the other, each thinner than the other. The simplest way to determine who is really thinner or younger in all these paradoxes is to bring the two twins together, which requires yanking one of the twins around, which in turn determines which twin is "really" moving.

Although these mind-bending paradoxes were indirectly resolved in Einstein's favor at the atomic level with studies of cosmic rays and atom smashers, this effect is so small that it was not directly seen in the laboratory until 1971, when airplanes carrying atomic clocks were sent into the air at great speeds. Because these atomic clocks can measure the passing of time with astronomical precision, scientists, by comparing the two clocks, could verify that time beat slower the faster you moved, exactly as Einstein had predicted.)

Another paradox involves two objects, each shorter than the other. Imagine a big-game hunter trying to trap a tiger about 10 feet long with a cage that is only 1 foot wide. Normally, this is impossible. Now imagine that the tiger is moving so fast that it shrinks to only 1 foot, so the cage can drop and capture the tiger. As the tiger screeches to a halt, it expands. If the cage is made of webbing, the tiger breaks the webbing. If the cage is made of concrete, the poor tiger is crushed to death.

But now look at the situation from the point of view of the tiger. If the tiger is at rest, the cage is now moving and has shrunk down to only one-tenth of a foot. How can a cage that small catch a tiger 10 feet long? The answer is that as the cage drops, it shrinks in the direction of motion, so it becomes a parallelogram, a squashed square. The two ends of the cage therefore do not necessarily hit the tiger simultaneously. What is simultaneous to the big-game hunter is not simultaneous to the tiger. If the cage is made of webbing, then the front part of the cage hits the tiger's nose first and begins to rip. As the cage drops, it continues to rip along the tiger's body, until the back end of the cage finally hits the tail. If the cage is made of concrete, then the tiger's nose is crushed first. As the cage descends, it continues to crush the length of the tiger's body, until the back end of the cage finally captures the tail.

These paradoxes even seized the public's imagination, with the following limerick running in the humor magazine *Punch*:

> *There once was a young lady named Bright*
> *Who could travel much faster than light*
> *She set out one day, in a relative way*
> *And came back the previous night.*

By this time, his good friend Marcel Grossman was a professor at the Polytechnic, and he sounded out Einstein to see if he wanted a position at his old school, this time as a full professor. Letters of recommendation spoke of Einstein in the highest terms. Marie Curie wrote that "mathematical physicists are unanimous in considering his work as being of the first rank."

So, just sixteen months after arriving in Prague, he returned to Zurich and the old Polytechnic. Returning to the Polytechnic (since 1911 called the Swiss Federal Institute of Technology, or ETH), this time as a famous professor, marked a personal victory for Einstein. He left the university with his name clouded in disgrace, with professors like Weber actively sabotaging his career. He returned as the leader of the new revolution in physics. That year, he received his first nomination for the Nobel Prize in physics. His ideas were still considered too radical for the Swedish academy, and there were dissident voices among Nobel laureates who wanted to sabotage his nomination. In 1912, the Nobel Prize did not go to Einstein, but to Nils Gustaf Dalén, for his work on improving lighthouses. (Ironically, lighthouses today have been made largely obsolete by the introduction of the global positioning satellite system, which depends crucially on Einstein's theory of relativity.)

Within another year, Einstein's reputation was growing so rapidly that he began to get inquiries from Berlin. Max Planck was eager to capture this rising star in physics, and Germany was the unquestioned leader of the world's research in physics, the crown jewel of German research being in Berlin. Einstein hesitated at first, since he had renounced his German citizenship and still nursed some bitter memories from his youth, but the offer was too tempting.

In 1913, Einstein was elected to the Prussian Academy of Sciences and later offered a position in Berlin at the university. He would be made director of the Kaiser Wilhelm Institute for Physics. But beyond the titles, which meant little to him, what made the offer especially attractive was the fact that there were no teaching obligations. (Although Einstein was a popular lecturer among the students, noted for treating his students with respect and kindness, teaching detracted from his main interest, general relativity.)

In 1914, Einstein arrived in Berlin to meet the faculty. He felt a bit nervous as they looked him over. Einstein would write, "The gentlemen in Berlin are gambling on me as if I were a prize hen. As for myself, I don't even know whether I'm going to lay another egg." The thirty-five-year-old rebel, with strange politics and stranger attire, soon had to adjust to the stiff, upper-crust ways of the Prussian Academy of Sciences, where members addressed each other as "Privy Councillor" or "Your Excellency." Einstein would muse, "It seems that most members confine themselves to displaying a peacock-like grandeur in writing; otherwise they are quite human."

Einstein's triumphant march from the patent office in Bern to the top ranks of German research was not without its personal toll. As his fame began to rise within the scientific community, his personal life began to unravel. These were Einstein's most productive years, bearing fruit that would eventually reshape human history, and almost impossible demands were being placed on his time, estranging him from his wife and children.

Einstein wrote that living with Mileva was like living in a cemetery, and when alone he tried to avoid being in the same room with her. His friends split on the question of who was mainly to blame. Many believed that Mileva was becoming increasingly isolated and resentful of her famous husband. Even Mileva's friends were distressed that she had aged considerably in those years and had let her appearance deteriorate noticeably. She was becoming increasingly shrill and cold, jealous even of the time he spent with his colleagues. When she discovered a letter of congratulations sent to Einstein by Anna Schmid (who knew Einstein during his brief time in Aarau and had since married), she blew her top, precipitating perhaps one of the angriest rifts in their already shaky marriage.

On the other hand, others believed that Einstein was hardly the perfect husband, constantly on the road, leaving Mileva to raise two children mainly by herself. Travel at the turn of the century was notoriously difficult, and extensive travels were taking him away for days and weeks. Like passing ships in the night, when he was home, they would only meet briefly at night for dinner or the theater. He was so immersed in the abstract world of mathematics that he had little emotional energy to connect with his wife. Worse, the more she complained to him about his absences, the more he withdrew into the world of physics.

It is probably safe to say that there is some truth in both allegations and it is pointless to assign blame. In retrospect, it was probably inevitable that the marriage would experience enormous strains. Perhaps their friends were right years ago when they said that the two were incompatible.

But the final break was precipitated by his acceptance of the offer from Berlin. Mileva was reluctant about going to Berlin. Perhaps being a Slav in the center of a Teutonic culture was too intimidating to her; more importantly, many of Einstein's relatives lived in Berlin, and Mileva feared being under their harsh, disapproving gaze. It was no secret that her in-laws hated her. At first, Mileva and the children made the trip to Berlin with Einstein, but then suddenly she left for Zurich, taking the children with her. They would never be united again. Einstein, who cherished his children more than anyone, was devastated. From that point on, he was forced to maintain a long-distance relationship with his sons, making the grueling ten-hour trip from Berlin to Zurich for visits. (When Mileva was eventually awarded custody of the children, Einstein's secretary, Helen Dukas, wrote that he cried all the way home.)

But what probably also precipitated the rupture was the growing presence of a certain cousin of Einstein's in Berlin. He would confess, "I live a very withdrawn life but not a lonely one, thanks to the care of a female cousin who actually drew me to Berlin in the first place."

Elsa Lowenthal was a double cousin; her mother and Einstein's mother were sisters, and their grandfathers were brothers. She was divorced, living with her two daughters, Margot and Ilse, just upstairs from her parents (Einstein's aunt and uncle). She and Einstein met briefly in 1912 when he visited Berlin. By then, Einstein had apparently decided that his marriage to Mileva was finished and that divorce was inevitable. However, he feared the repercussions a divorce would have on his young sons.

Ever since they were children, Elsa had taken a liking to Einstein. She confessed to having fallen in love with him as a child when she heard him play Mozart. But what apparently most attracted her was his rising stardom in the academic world, his respect by physicists around the world. In fact, she made it no secret that she loved to bask in this fame. Like Mileva, she was older, four years older than Einstein. But that is where the resemblance ended. In fact, they were like polar opposites.

Einstein, in fleeing Mileva, was apparently going overboard in the other direction. While Mileva was often uncaring of her appearance and looked continually harassed, Elsa was highly bourgeois and conscious of class ranking. She was always trying to cultivate acquaintances in intellectual circles in Berlin and would proudly show off Einstein to all her friends in high society. Unlike Mileva, who was laconic, withdrawn, and moody, Elsa was a social butterfly, fluttering between dinner parties and theater openings. And unlike Mileva, who gave up trying to reform her husband, Elsa was more of a mother, continually correcting his manners while devoting her full energies to helping him fulfill his destiny. A Russian journalist later summed up the relationship between Einstein and Elsa: "She is all love for her great husband, always ready to shield him from the harsh intrusions of life and to ensure the peace of mind necessary for his great ideas to mature. She is filled with the realization of his great purpose as a thinker and with the tenderest feelings of companion, wife, and mother towards a remarkable, exquisite, grown-up child."

After Mileva stormed out of Berlin in 1915, taking the children with her, Einstein and Elsa got even closer. What consumed Einstein during this important period, however, was not love, but the universe itself.

PART II
SECOND PICTURE
Warped Space-Time

CHAPTER 4

General Relativity and "the Happiest Thought of My Life"

E instein was still not satisfied. He was already ranked among the top physicists of his time, yet he was restless. He realized that there were at least two glaring holes in his theory of relativity. First, it was based entirely on inertial motions. In nature, however, almost nothing is inertial. Everything is in a state of constant acceleration: the jostling of trains, the zigzags of falling leaves, the rotation of the earth around the sun, the motion of heavenly bodies. Relativity theory failed to account for even the commonest acceleration found on the earth.

Second, the theory said nothing about gravity. It made the sweeping claim that it was a universal symmetry of nature, applying to all sectors of the universe, yet gravity seemed beyond its reach. This was also quite embarrassing, because gravity is everywhere. The deficiencies of relativity were obvious. Since the speed of light was the ultimate speed of the universe, relativity theory said that it would take eight minutes for any disturbance on the sun to reach the earth. This, however, contradicted Newton's theory of gravity, which stated that gravitational effects were instantaneous. (The speed of Newton's gravity was infinite, since the speed of light does not appear anywhere in Newton's equations.) Einstein therefore needed to completely overhaul Newton's equations to incorporate the speed of light.

In short, Einstein realized the immensity of the problem of generalizing his relativity theory to include accelerations and gravity. He began to call his earlier theory of 1905 the "special theory of relativity," to differentiate it from the more powerful "general theory of relativity" that was needed to describe gravity. When he told Max Planck of his ambitious program, Planck warned him, "As an older friend, I must advise you against it for in the first place you will not succeed, and even if you succeed, no one will believe you." But Planck also realized the importance of the problem when he said, "If you are successful, you will be called the next Copernicus."

The key insight into a new theory of gravity took place while Einstein was still slaving over patent applications as a lowly civil servant back in 1907. He would recall, "I was sitting in a chair in the patent office at Bern when all of a sudden, a thought occurred to me: If a person falls freely, he will not feel his own weight. I was startled. This simple thought made a deep impression on me. It impelled me toward a theory of gravitation."

In an instant, Einstein realized that if he had fallen over off his chair, he would be momentarily weightless. For example, if you are in an elevator and the cable suddenly breaks, you would be in free fall; you would fall at the same rate as the elevator floor. Since both you and the elevator are now falling at the same speed, it would appear that you are weightless, floating in air. Similarly, Einstein realized that if he fell over off his chair, he would be in free fall and the effect of gravity would be canceled perfectly by his acceleration, making him appear weightless.

This concept is an old one. It was known to Galileo, who in an apocryphal story dropped a small rock and a large cannonball from the Leaning Tower of Pisa. He was the first to show that all objects on Earth accelerate at precisely the same rate under gravity (32 feet per second squared). Newton also knew this fact when he realized that the planets and the moon were actually in a state of free fall in their orbit around the sun or the earth. Every astronaut who has ever been shot into outer space also realizes that gravity can be canceled by acceleration. In a rocket ship, everything inside, including the floor, the instruments, and you, falls at the same rate. Thus, when you look around, everything is floating. Your feet drift above the floor, giving the illusion that gravity has vanished, because the floor is falling along with your body. And if an astronaut takes a space walk outside the ship, he or she does not suddenly drop to the earth, but instead floats gently alongside the rocket because both the rocket and the astronaut are falling in unison even as they orbit the earth. (Gravity has not actually disappeared in outer space, as many science books erroneously claim. The sun's gravity is powerful enough to whip the planet Pluto in its orbit billions of miles from Earth. Gravity has not disappeared; it has just been canceled by the falling of the rocket ship beneath your feet.)

This is called the "equivalence principle," in which all masses fall at the same rate under gravity (more precisely, the inertial mass is the same as the gravitational mass). This was indeed an old idea, almost a curiosity to Galileo and Newton, but in the hands of a seasoned physicist like Einstein, it was to become the foundation of a new relativistic theory of gravity. Einstein went one giant step further than Galileo or Newton. He formulated his next postulate, the postulate behind general relativity: *The laws of physics in an accelerating frame or a gravitating frame are indistinguishable.* Remarkably, this simple statement, in the hands of Einstein, became the basis of a theory that would give us warped space, black holes, and the creation of the universe.

After this brilliant insight of 1907 in the patent office, it took years for Einstein's new theory of gravity to gestate. A new picture of gravity was emerging from the equivalence principle, but it wouldn't be until 1911 that he began to publish the fruits of his thoughts. The first consequence of the equivalence principle is the fact that light must bend under gravity. The idea that gravity might influence light beams is an old one, dating back at least to the time of Isaac Newton. In his book *Opticks*, he asked whether or not gravity can influence starlight: "Do not Bodies act upon light at a Distance, and by their Action bend its Rays; and is not this Action strongest at the least Distance?" Unfortunately, given the technology of the seventeenth century, he could give no answer.

But now Einstein, after more than two hundred years, returned to this question. Consider turning on a flashlight inside a rocket ship that is accelerating in outer space. Because the rocket is accelerating upward, the light beam droops downward. Now invoke the equivalence principle. Since the physics inside the spaceship must be indistinguishable from the physics on Earth, it means that *gravity must also bend light.* In a few brief steps, Einstein was led to a new physical phenomenon, the bending of light due to gravity. He immediately realized that such an effect was calculable.

The largest gravitational field in the solar system is generated by the sun, so Einstein asked himself whether the sun was sufficient to bend starlight from distant stars. This could be tested by taking two photographs of the same collection of stars in the sky at two different seasons. The first photo of these stars would be taken at night when starlight is undisturbed; the second photo would be taken several months later when the sun is positioned directly in front of this same collection of stars. By comparing the two photographs, one might be able to measure how the stars have shifted slightly in the sun's vicinity due to the sun's gravity. Because the sun overwhelms the light coming from the stars, any experiment on the bending of starlight would have to be performed during a solar eclipse, when the moon blocks out the light from the sun and the stars become visible during the daytime. Einstein reasoned that photographs of the day sky taken during an eclipse, compared to photographs taken of the same sky at night, should show a slight distortion in the location of the stars in the vicinity of the sun. (The presence of the moon also bends starlight a bit because of the moon's gravity, but this is a very tiny amount compared to the bending of starlight caused by the sun, which is much larger. Thus, the bending of starlight during an eclipse is not affected by the presence of the moon.)

The equivalence principle could help him to calculate the approximate motion of light beams as they were pulled by gravity, but it still did not tell him anything about gravity itself. What was lacking was a *field theory of gravity*. Recall that Maxwell's equations describe a genuine field theory, in which lines of force are like a spider web that could vibrate and support waves traveling along the lines of force. Einstein sought a gravitational field whose lines of force could support gravitational vibrations that traveled at the speed of light.

Around 1912, after years of concentrated thought, he slowly began to realize that he needed to overhaul our understanding of space and time; to do so required new geometries beyond those inherited from the ancient Greeks. The key observation that sent him on the road to curved space-time was a paradox, sometimes referred to as "Ehrenfest's paradox," that his friend Paul Ehrenfest once posed to Einstein. Consider a simple merry-go-round or a spinning disk. At rest, we know that its circumference is equal to π times the diameter. Once the merry-go-round is set into motion, however, the outer rim travels faster than the interior and hence, according to relativity, should shrink more than the interior, distorting the shape of the merry-go-round. This means that the circumference has shrunk and is now less than π times the diameter; that is, the surface is no longer flat. *Space is curved.* The surface of the merry-go-round can be compared to the area within the Arctic Circle. We can measure the diameter of the Arctic Circle by walking from one point on the circle, directly across the North Pole, to the opposite point on the circle. Then we can measure the circumference of the Arctic Circle. If we compare the two, we also find that the circumference is less than π times the diameter because the earth's surface is curved. But for the last two thousand years, physicists and mathematicians relied on Euclidean geometry, which is based on *flat* surfaces. What would happen if they imagined a geometry based on *curved* surfaces?

Once we realize that space can be curved, a startling new picture emerges. Think of a heavy rock placed on a bed. The rock, of course, will sink into the bed. Now shoot a tiny marble over the bed. The marble will not move in a straight line but in a curved line around the rock. There are two ways to analyze this effect. From a distance, a Newtonian may say that there is a mysterious "force" that emanates from the rock to the marble, forcing the marble to change its path. This force, although invisible, reaches out and pulls on the marble. However, a relativist may see an entirely different picture. To a relativist looking at the bed close up, there is no force that pulls the marble. There is just the depression in the bed, which dictates the motion of the marble. As the marble moves, the surface of the bed "pushes" the marble until it moves in a circular motion.

Now replace the rock with the sun, the marble with the earth, and the bed with space and time. Newton would say that an invisible force called "gravity" pulls the earth around the sun. Einstein would reply that there is no gravitational pull at all. The earth is deflected around the sun because the curvature of *space itself* is pushing the earth. In a sense, gravity does not pull, but space pushes.

In this picture, Einstein could explain why it would take eight minutes for any disturbance on the sun to reach the earth. For example, if we suddenly remove the rock, the bed will spring back to normal, creating ripples that travel at a definite speed across the bed. Similarly, if the sun were to disappear, it would create a shock wave of warped space that would travel at the speed of light. This picture was so simple and elegant that he could explain the essential idea to his second son, Eduard, who asked him why he was so famous. Einstein replied, "When a blind beetle crawls over the surface of a curved branch, it doesn't notice that the track it has covered is indeed curved. I was lucky enough to notice what the beetle didn't notice."

Newton, in his landmark *Philosophiae Naturalis Principia Mathematica*, confessed that he was unable to explain the origin of this mysterious pull, which acted instantly throughout the universe. He coined his famous phrase *hypotheses non fingo* (I frame no hypotheses) because of his inability to explain where gravity came from. With Einstein, we see that gravity is caused by the bending of space and time. "Force" is now revealed to be an illusion, a by-product of geometry. In this picture, the reason why we are standing on the earth is not because the earth's gravity pulls us down. According to Einstein, there is no gravitational pull. The earth warps the space-time continuum around our bodies, so space itself pushes us down to the floor. Thus, it is the presence of matter that warps space around it, giving us the illusion that there is a gravitational force pulling on neighboring objects.

This bending, of course, is invisible, and from a distance, Newton's picture appears to be correct. Think of ants walking on a crumpled sheet of paper. Trying to follow a straight line, they find that they are constantly being tugged to the left and right as they walk over the folds in the paper. To the ants, it appears as if there is a mysterious force pulling them in both directions. However, to someone looking down on the ants, it is obvious that there is no force, there is just the bending of the paper pushing on the ants, which gives the illusion that there is a force. Recall that Newton thought of space and time as an absolute reference frame for all motions. However, to Einstein, space and time could assume a dynamic role. If space is curved, then anyone moving on this stage would think that mysterious forces were acting on their bodies, pushing them one way or the other.

By comparing space-time to a fabric that can stretch and bend, Einstein was forced to study the mathematics of curved surfaces. He quickly found himself buried in a morass of mathematics, unable to find the right tools to analyze his new picture of gravity. In some sense, Einstein, who once scorned mathematics as "superfluous erudition," was now paying for the years in which he cut the mathematics courses at the Polytechnic.

In desperation, he turned to his friend, Marcel Grossman. "Grossman, you must help me or else I'll go crazy!" Einstein confessed, "Never in my life have I tormented myself anything like this, and that I have become imbued with a great respect for mathematics, the more subtle parts of which I had previously regarded as sheer luxury! Compared to this problem the original relativity theory is child's play."

When Grossman reviewed the mathematical literature, he found that, ironically enough, the basic mathematics that Einstein needed had indeed been taught at the Polytechnic. In the geometry of Bernhard Riemann, developed in 1854, Einstein finally found the mathematics powerful enough to describe the bending of space-time. (Years later, when looking back at how difficult it was to master new mathematics, Einstein noted to some junior high school students, "Do not worry about your difficulties in mathematics; I can assure you that mine are still greater.")

Before Riemann, mathematics was based on Euclidean geometry, the geometry of flat surfaces. Schoolchildren for thousands of years had been grilled in the time-honored theorems of Greek geometry, where the sum of the interior angles of a triangle equals 180 degrees, and parallel lines never meet. Two mathematicians, the Russian Nicolai Lobachevsky and the Austro-Hungarian János Bolyai, came extremely close to developing a non-Euclidean geometry, that is, in which the sum of the angles of a triangle can be more or less than 180 degrees. But the theory of non-Euclidean geometry was finally developed by the "prince of mathematics," Carl Friedrich Gauss, and especially his student, Riemann. (Gauss suspected that Euclid's theory might be incorrect even on physical grounds. He had his assistants shine light beams from atop the Harz Mountains, trying to experimentally calculate the sum of the angles of a triangle formed by three mountaintops. Unfortunately, he got a negative result. Gauss was also such a politically cautious individual that he never published his work on this sensitive subject, fearing the ire of conservatives who swore by Euclidean geometry.)

Riemann discovered entirely new worlds of mathematics—the geometry of curved surfaces in any dimension, not just two or three spatial dimensions. Einstein was convinced these higher geometries would yield a more accurate description of the universe. For the first time, the mathematical language of "differential geometry" was working its way into the world of physics. Differential geometry, or tensor calculus, the mathematics of curved surfaces in any dimension, was once considered to be the most "useless" branch of mathematics, devoid of any physical content. Suddenly, it was transformed into the language of the universe itself.

In most biographies, Einstein's theory of general relativity is presented as fully developed in 1915, as if he unerringly found the theory fully formed by magic. However, only in the last decades have some of Einstein's "lost notebooks" been analyzed, and they fill in the many missing gaps between 1912 and 1915. Now it is possible to construct, sometimes month by month, the crucial evolution of one of the greatest theories of all time. In particular, he wanted to generalize the notion of covariance. Special relativity, as we saw, was based on the idea of Lorentz covariance, that is, that the equations of physics retain the same form under a Lorentz transformation. Now Einstein wanted to generalize this to all possible accelerations and transformations, not just inertial ones. In other words, he wanted equations that retained the same form no matter what frame of reference was used, whether it was accelerating or moving with constant velocity. Each frame of reference in turn requires a coordinate system to measure the three dimensions of space and the time. What Einstein desired was a theory that retained its form no matter which distance and time coordinates were used to measure the frame. This led him to his famed principle of general covariance: *the equations of physics must be generally covariant* (i.e., they must maintain the same form under an arbitrary change of coordinates).

For example, think of throwing a fishing net over a tabletop. The fishing net represents an arbitrary coordinate system, and the area of the tabletop represents something that remains the same under any distortion of the fishing net. No matter how we twist or curl up the fishing net, the area of the table top remains the same.

In 1912, aware that Riemann's mathematics was the correct language for gravitation, and guided by the law of general covariance, Einstein searched within Riemannian geometry for objects that are generally covariant. Surprisingly, there were only two covariant objects available to him: the volume of a curved space and the curvature (called the "Ricci curvature") of such a space. This was of immense help: by severely restricting the possible building blocks used to construct a theory of gravity, the principle of general covariance led Einstein to formulate the essentially correct theory in 1912, after only a few months of examining Riemann's work, based on the Ricci curvature. For some reason, however, he threw away the correct theory of 1912 and began to pursue an incorrect idea. Precisely why he abandoned the correct theory was a mystery to historians until recently, when the lost notebooks were discovered. That year, when he essentially constructed the correct theory of gravity out of the Ricci curvature, he made a crucial mistake. He thought that this correct theory violated what is known as "Mach's principle." One particular version of this principle postulates that the presence of matter and energy in the universe uniquely determines the gravitational field surrounding it. Once you fix a certain configuration of planets and stars, then the gravitation surrounding these planets and stars is fixed. Think, for example, of throwing a pebble into a pond. The larger the pebble, the greater the ripples on the pond. Thus, once we know the precise size of the pebble, the distortion of the pond can be uniquely determined. Likewise, if we know the mass of the sun, we can uniquely determine the gravitational field surrounding the sun.

This is where Einstein made his mistake. He thought that the theory based on the Ricci curvature violated Mach's principle because the presence of matter and energy did not uniquely specify the gravitational field surrounding it. With his friend Marcel Grossman, he tried to develop a more modest theory, one that was covariant just under rotations (but not general accelerations). Because he abandoned the principle of covariance, however, there was no clear path to guide him, and he spent three frustrating years wandering in the wilderness of the Einstein-Grossman theory, which was neither elegant nor useful—for instance, it failed to yield Newton's equations for small gravitational fields. Although Einstein had perhaps the best physical instincts of anyone on Earth, he ignored them.

While groping for the final equations, Einstein focused on three key experiments that might prove his ideas concerning curved space and gravity: the bending of starlight during an eclipse, the red shift, and the perihelion of Mercury. In 1911, even before his work on curved space, Einstein held out hope that an expedition could be sent to Siberia during the solar eclipse of August 21, 1914, to find the bending of starlight by the sun.

The astronomer Erwin Finlay Freundlich was to investigate this eclipse. And Einstein was so convinced of the correctness of his work that at first he offered to fund the ambitious project out of his own pocket. "If everything fails, I'll pay for the thing out of my own slight savings, at least the first 2,000 marks," he wrote. Eventually, a wealthy industrialist agreed to provide the funding. Freundlich left for Siberia a month before the solar eclipse, but Germany declared war on Russia, and he and his assistant were taken prisoner and their equipment was confiscated. (In hindsight, perhaps it was fortunate for Einstein that the 1914 expedition was unsuccessful. If the experiment had been performed, the results would not, of course, have agreed with the value predicted by Einstein's incorrect theory, and his entire program might have been disgraced.)

Next, Einstein calculated how gravity would affect the frequency of a light beam. If a rocket is launched from the earth and sent into outer space, the gravity of the earth acts like a drag, pulling the rocket back. Energy is therefore lost as the rocket struggles against the pull of gravity. Similarly, Einstein reasoned that if light were emitted from the sun, then gravity would also act as a drag on the light beam, making it lose energy. The light beam will not change in velocity, but the frequency of the wave will drop as it loses energy struggling against the sun's gravity. Thus, yellow light from the sun will decrease in frequency and become redder as the light beam leaves the sun's gravitational pull. Gravitational red shift, however, is an extremely small effect, and Einstein had no illusion that it would be tested in the laboratory any time soon. (In fact it would take four more decades before gravitational red shift could be seen in the laboratory.)

Last, he set out to solve an age-old problem: why the orbit of Mercury wobbles and deviates slightly from Newton's laws. Normally, the planets execute perfect ellipses in their journeys around the sun, except for slight disturbances caused by the gravity of nearby planets, which results in a trajectory resembling the petals of a daisy. The orbit of Mercury, however, even after subtracting the interference caused by nearby planets, showed a small but distinct deviation from Newton's laws. This deviation, called the "perihelion," was first observed in 1859 by astronomer Urbain Leverrier, who calculated a tiny shift of 43.5 seconds of arc per century that could not be explained by Newton's laws. (The fact that there were apparent discrepancies in Newton's laws of motion was not new. In the early 1800s, when astronomers were puzzled by a similar wobbling of the orbit of Uranus, they faced a stark choice: either abandon the laws of motion or postulate that there was another unknown planet tugging on the orbit of Uranus. Physicists breathed a sigh of relief when in 1846, a new planet, christened Neptune, was discovered just where Newton's laws predicted it should be.)

But Mercury was the remaining puzzle. Rather than discard Newton, astronomers in time-honored tradition postulated the existence of a new planet called "Vulcan," circling the sun within the orbit of Mercury. In repeated searches of the night sky, however, astronomers could find no experimental evidence for such a planet.

Einstein was prepared to accept the more radical interpretation: perhaps Newton's laws themselves were incorrect, or at least incomplete. In November 1915, after wasting three years on the Einstein-Grossman theory, he went back to the Ricci curvature, which he had discarded back in 1912, and spotted his key mistake. (Einstein had dropped the Ricci curvature because it yielded more than one gravitational field generated by a piece of matter, in seeming violation of Mach's principle. But then, because of general covariance, he now realized that these gravitational fields were actually mathematically equivalent and yielded the same physical result. This impressed upon Einstein the power of general covariance: not only did it severely restrict the possible theories of gravity, it also yielded unique physical results because many gravitational solutions were equivalent.)

In perhaps the greatest mental concentration of Einstein's life, he then slaved away at his final equation, shutting out all distractions and working himself mercilessly to see if he could derive the perihelion of Mercury. His lost notebooks show that he would repeatedly propose a solution and then ruthlessly check to see that it reproduced Newton's old theory in the limit of small gravitational fields. This task was extremely tedious, since his tensor equations consisted of ten distinct equations, rather than the single equation of Newton. If it failed, then he would try another solution to see if that reproduced Newton's equation. This exhaustive, almost Herculean task was finally completed in late November 1915, leaving Einstein totally drained. After a long tedious calculation with his old theory of 1912, he found that it predicted the deviation in Mercury's orbit to be 42.9 seconds of arc per century, well within acceptable experimental limits. Einstein was shocked beyond belief. This was exhilarating, the first solid experimental evidence that his new theory was correct. "For some days, I was beyond myself with excitement," he recalled. "My boldest dreams have now come true." The dream of a lifetime, to find the relativistic equations for gravity, was realized.

What thrilled Einstein was that through the abstract physical and mathematical principle of general covariance, he could derive a solid, decisive experimental result: "Imagine my joy over the practicability of general covariance and over the result that the equations correctly yield the perihelion movement of Mercury." With the new theory, he then recalculated the bending of starlight by the sun. The addition of curved space to his theory meant that this final answer was 1.7 seconds of arc, twice his original value (about 1/2000th of a degree).

He was convinced that the theory was so simple, elegant, and powerful that no physicist could escape its hypnotic spell. "Hardly anyone who has truly understood it will be able to escape the charm of this theory," he would write. "The theory is of incomparable beauty." Miraculously, the principle of general covariance was so powerful a tool that the final equation, which would describe the structure of the universe itself, was only 1 inch long. (Physicists today marvel that an equation so short can reproduce the creation and evolution of the universe. Physicist Victor Weisskopf likened that sense of wonder to the story of a peasant who saw a tractor for the first time in his life. After examining the tractor and peering under the hood, he asks in bewilderment, "But where is the horse?")

The only thing to mar Einstein's triumph was a minor priority fight with David Hilbert, perhaps the world's greatest living mathematician. While the theory was in its last, final steps before completion, Einstein had given a series of six two-hour lectures at Göttingen for Hilbert. Einstein still lacked certain mathematical tools (called the "Bianchi identities") that prevented him from deriving his equations from a simple form, called the "action." Later, Hilbert filled in the final step in the calculation, wrote down the action, and then published the final result by himself, just six days ahead of Einstein. Einstein was not pleased. In fact, he believed that Hilbert had tried to steal the theory of general relativity by filling in the final step and taking credit. Eventually, the rift between Einstein and Hilbert healed, but Einstein became wary of sharing his results too freely. (Today, the action by which one derives general relativity is known as the "Einstein-Hilbert action." Hilbert was probably led to finish the last tiny piece of Einstein's theory because, as he often said, "physics is too important to be left to the physicists"; that is, physicists probably were not mathematically skilled enough to probe nature. This view apparently was shared by other mathematicians. The mathematician Felix Klein would grumble that Einstein was not innately a mathematician, but worked under the influence of obscure physical-philosophical impulses. That is probably the essential difference between mathematicians and physicists and why the former have consistently failed to find new laws of physics. Mathematicians deal exclusively with scores of small, self-consistent domains, like isolated provinces. Physicists, however, deal with a handful of simple physical principles that may require many mathematical systems to solve. Although the language of nature is mathematics, the driving force behind nature seems to be these physical principles, for example, relativity and the quantum theory.)

News of Einstein's new theory of gravity was interrupted by the outbreak of war. In 1914, the assassination of an obscure Yugoslavian archduke touched off the greatest bloodletting of its time, drawing the British, Austro-Hungarian, Russian, and Prussian empires into a catastrophic conflict that would doom tens of millions of young men. Almost overnight, quiet, distinguished professors at German universities became bloodthirsty nationalists. Nearly the entire faculty at the University of Berlin was swept up by the war fever and devoted all their energies to the war effort. In support of the Kaiser, ninety-three prominent intellectuals signed the notorious "Manifesto to the Civilized World," which called for all people to rally around the Kaiser and ominously declared that the German people must defy "Russian hordes allied with Mongols and Negroes unleashed against the white race." The manifesto justified the German invasion of Belgium and proudly proclaimed, "The German Army and the German people are one. This awareness now binds seventy million Germans without distinction of education, class, or party." Even Einstein's benefactor, Max Planck, signed the manifesto, as did such distinguished individuals as Felix Klein and physicists Wilhelm Roentgen (the discoverer of X-rays), Walther Nernst, and Wilhelm Ostwald.

Einstein, a confirmed pacifist, refused to sign the manifesto. Georg Nicolai, Elsa's physician, was a prominent anti-war activist and asked one hundred intellectuals to sign a counter-manifesto. Because of the overwhelming war hysteria gripping Germany, only four actually did sign it, among them Einstein. Einstein could only shake his head in disbelief, writing,

"Unbelievable what Europe has unleashed in its folly." He added sadly, "At such a time as this one realizes what a sorry species of animal one belongs to."

In 1916, Einstein's world was rocked once again, this time by the astonishing news that his close idealistic friend, Friedrich Adler, the same physicist who generously gave up a potential professorship at the University of Zurich in Einstein's favor, had assassinated the Austrian prime minister, Count Karl von Stürgkh, in a crowded Vienna restaurant, shouting, "Down with tyranny! We want peace!" The entire country was riveted by the news that the son of the founder of the Austrian Social Democrats had committed an unspeakable act of murder against the nation. Adler was immediately sent to prison, where he faced a possible death sentence. While awaiting his trial, Adler returned to his favorite pastime, physics, and began writing a long essay that was critical of Einstein's theory of relativity. In fact, in the midst of all the turmoil he had created by the assassination and its potential consequences, he preoccupied himself with the idea that he had found a crucial error in relativity!

Adler's father Viktor seized upon the only possible defense available to his son. Realizing that mental illness ran in the family, Viktor stated that his son was mentally deranged, and pleaded for leniency. As proof of his madness, Viktor pointed to the fact that his son was trying to disprove Einstein's well-accepted theory of relativity. Einstein offered to be a character witness, but he was never called.

Though the court originally found Adler guilty and sentenced him to death by hanging, the sentence was later changed to life imprisonment as a result of pleas on his behalf by Einstein and others. (Ironically, with the subsequent collapse of the government after World War I, Adler was freed in 1918 and was even elected to the Austrian National Assembly, becoming one of the most popular figures in the labor movement.)

The war and the great mental effort necessary to create general relativity inevitably took a toll on Einstein's health, which was always precarious. He finally collapsed in pain in 1917, suffering from a near breakdown. So weakened was he by his Herculean mental feat that he was unable to leave his apartment. His weight plummeted dangerously by 56 pounds in just two months. Becoming a shell of his former self, he felt that he was dying of cancer, but was diagnosed with a stomach ulcer. The physician recommended complete rest and a change in diet. During this period, Elsa became a constant companion, nursing the ailing Einstein slowly back to health. He grew much closer to Elsa and her daughters as well, especially after he moved into an apartment next to hers.

In June 1919, Einstein finally married Elsa. With very definite ideas of what a distinguished professor should dress like, she helped to usher in his transition from a bohemian, bachelor professor to an elegant, domesticated husband, perhaps preparing him for the next development in his life as he emerged a heroic figure on the world stage.

CHAPTER 5

The New Copernicus

Einstein, recovering from the disruption and chaos of World War I, eagerly awaited the analysis of the next solar eclipse, to take place on May 29, 1919. One British scientist, Arthur Eddington, was keenly interested in performing the decisive experiment to test Einstein's theory. Eddington was secretary of the Royal Astronomical Society in England and was equally at ease with performing astronomical observations by telescope and delving into the mathematics of general relativity. He also had another reason for performing the solar eclipse experiment: he was a Quaker, and his pacifist beliefs prevented him from fighting with the British army during World War I. In fact, he was fully prepared to go to prison rather than be inducted into the military. The officials of Cambridge University feared a scandal if one of its young stars went to jail as a conscientious objector, so they were able to negotiate a deferment from the government, on the stipulation that he perform a civic duty, specifically, leading an expedition to observe the solar eclipse of 1919 and test Einstein's theory. So now it was his official patriotic duty for the war effort to lead the expedition to test general relativity.

Arthur Eddington set up camp at the island of Principe, in the Gulf of Guinea, off the coast of West Africa, and another team, led by Andrew Crommelin, set sail to Sobral in northern Brazil. Bad weather conditions, with rain clouds blocking out the sun, almost ruined the entire experiment. But the clouds miraculously parted just enough for photographs to be taken of the stars at 1:30 in the afternoon.

It would be months, however, before the teams could return to England and carefully analyze their data. When Eddington finally compared his photographs with other photographs taken in England several months earlier with the same telescope, he found an average deflection of 1.61 arc seconds, while the Sobral team determined a value of 1.98 arc seconds. Taking an average, they calculated 1.79 arc seconds, which confirmed Einstein's prediction of 1.74 arc seconds to within experimental error. Eddington would later fondly recall that verifying Einstein's theory was the greatest moment in his life.

On September 22, 1919, Einstein finally received a cable from Hendrik Lorentz, informing him of the fantastic news. Einstein excitedly wrote to his mother, "Dear Mother—Good news today. H. A. Lorentz cabled me that the English expedition really has proved the deflection of light by the sun." Max Planck apparently stayed up all night to see if the solar eclipse data would verify general relativity. Einstein joked later, "If he had *really* understood the general theory of relativity, he would have gone to bed the way I did."

Although the scientific community was now buzzing with the startling news of Einstein's new theory of gravity, the firestorm did not break publicly until a joint meeting of the Royal Society and the Royal Astronomical Society in London on November 6, 1919. Einstein was suddenly transformed from a senior, distinguished professor of physics in Berlin to a world figure, a worthy successor to Isaac Newton. At that meeting, philosopher Alfred Whitehead noted, "There was an atmosphere of tense interest that was exactly that of a Greek drama." Sir Frank Dyson was the first to speak. He said, "After careful study of the plates I am prepared to say that there can be no doubt that they confirm Einstein's prediction. A very definite result has been obtained that light is deflected in accordance with Einstein's law of gravity." The Nobel laureate J. J. Thomson, president of the Royal Society, said solemnly, this is "one of the greatest achievements in the history of human thought. It is not the discovery of an outlying island but of a whole continent of new scientific ideas. It is the greatest discovery in connection with gravitation since Newton enunciated his principles."

According to legend, as Eddington left the assembly, another scientist stopped him and asked, "There's a rumor that only three people in the entire world understand Einstein's theory. You must be one of them." Eddington stood in silence, so the scientist said, "Don't be modest Eddington." Eddington shrugged, and said, "Not at all. I was wondering who the third might be."

The next day, the *Times* of London splashed the headline: "Revolution in Science—New Theory of the Universe—Newton's Ideas Overthrown—Momentous Pronouncement—Space 'Warped.'" (Eddington wrote to Einstein, "All England is talking about your theory. . . . For scientific relations between England and Germany, this is the best thing that could have happened." The London newspapers also noted, approvingly, that Einstein did not sign the infamous manifesto of ninety-three German intellectuals that had infuriated British intellectuals.)

Eddington, in fact, would serve as Einstein's main proponent and keeper of the flame in the English-speaking world, defending general relativity against all challengers. Like Thomas Huxley in the previous century, who served as "Darwin's bulldog" to promote the heretical theory of evolution to a deeply religious Victorian England, Eddington would use the full force of his scientific reputation and considerable debating skills to promote relativity. This strange union between two pacifists, a Quaker and a Jew, helped to bring relativity to the English-speaking people.

So suddenly did this story burst on the world media that

many newspapers were caught off guard, scrambling to find anyone with a knowledge of physics. The *New York Times* hurriedly sent its golf expert, Henry Crouch, to cover this fast-breaking story, adding numerous errors in the process. The *Manchester Guardian* sent its music critic to cover the story. Later, the *Times* of London asked Einstein to elaborate on his new theory in an article. To illustrate the relativity principle, he wrote in the *Times*: "Today in Germany I am called a German man of science, and in England I am represented as a Swiss Jew. If I come to be regarded as a *bête noire*, the descriptions will be reversed, and I shall become a Swiss Jew for the Germans and a German man of science for the English."

Soon, hundreds of newspapers were clamoring for an exclusive interview with this certified genius, this successor to Copernicus and Newton. Einstein was besieged by reporters eager to make their deadlines. It seemed that every newspaper in the world was carrying this story on its front pages. Perhaps the public, exhausted by the carnage and senseless savagery of World War I, was ready for a mythic figure who tapped into their deepest myths and legends about the stars in the heavens, whose mystery has forever been in their dreams. Einstein, moreover, had redefined the image of genius itself. Instead of an aloof figure, the public was delighted that this messenger from the stars was a young Beethoven, complete with flaming hair and rumpled clothes, who could wisecrack with the press and thrill the crowds with learned one-liners and quips.

He wrote to his friends, "At present, every coachman and every waiter argues about whether or not the relativity theory is correct. A person's conviction on this point depends on the political party he belongs to." But after the novelty wore off, he began to see the down side to this publicity. "Since the flood of newspaper articles," he wrote, "I have been so swamped with questions, invitations, challenges, that I dream I am burning in Hell and the postman is the Devil eternally roaring at me, throwing new bundles of letters at my head because I have not answered the old ones." He concluded, "This world is a curious madhouse" with him at the center of this "relativity circus," as he called it. He lamented, "I feel now something like a whore. Everybody wants to know what I am doing." Curiosity seekers, cranks, circus promoters, all clamored for a piece of Albert Einstein. The *Berliner Illustrite Zeitung* detailed some of the problems faced by the suddenly famous scientist, who declined a generous offer from the London Palladium booking agent to include him on a bill with comedians, tightrope walkers, and fire eaters. Einstein could always politely say no to offers that would make him into a curiosity, but he could do nothing to prevent babies and even cigar brands from being named after him.

Anything as momentous as Einstein's discovery inevitably invited armies of skeptics to mount a counterattack. The skeptics were led by the *New York Times*. After recovering from the initial shock of being scooped by the British press, its editors kidded the British people for being so gullible, for being so quick to accept the theories of Einstein. The *New York Times* wrote that the British "seem to have been seized with something like intellectual panic when they heard of photographic verification of the Einstein theory. . . . They are slowly recovering as they realize that the sun still rises—apparently—in the east." What particularly irked the editors in New York and aroused their suspicion was that so very few people in the world could make any sense of the theory. The editors wailed that this was bordering on being un-American and undemocratic. Was the world being duped by a practical joker?

In the academic world, the skeptics were legitimized by a Columbia University professor of celestial mechanics, Charles Lane Poor. He mistakenly led the charge by stating, "The supposed astronomical proofs of the theory, as cited and claimed by Einstein, do not exist." Poor compared the author of relativity theory to the characters of Lewis Carroll: "I have read various articles on the fourth dimension, the relativity theory of Einstein, and other psychological speculation on the constitution of the universe; and after reading them I feel as Senator Brandegee felt after a celebrated dinner in Washington. 'I feel,' he said, 'as if I had been wandering with Alice in Wonderland and had tea with the Mad Hatter.' " Engineer George Francis Gillette fumed that relativity was "cross-eyed physics . . . utterly mad . . . the moronic brain child of mental colic . . . the nadir of pure drivel . . . and voodoo nonsense. By 1940, relativity will be considered a joke. Einstein is already dead and buried alongside Anderson, Grimm, and the Mad Hatter." Ironically, the only reason why historians still remember these individuals is their futile tirades against relativity theory. It is the hallmark of good science that physics is not determined by a popularity contest or by *New York Times* editorials, but by careful experimentation. As Max Planck once said, referring to the ferocious criticism that he once faced when proposing his quantum theory, "A new scientific truth does not as a rule prevail because its opponents declare themselves persuaded or convinced, but because the opponents gradually die out and the younger generation is made familiar with the truth from the start." Einstein himself once remarked, "Great spirits have always encountered violent opposition from mediocre minds."

Unfortunately, the adulation of Einstein in the press stimulated the hatred, jealousy, and bigotry of the growing army of his detractors. The most notorious hater of Jews in the physics establishment was Philipp Lenard, the Nobel Prize–winning physicist who had established the basic frequency dependence of the photoelectric effect, a result that was finally explained by Einstein's theory of the light quantum, the photon. Mileva had even attended Lenard's lectures when she visited Heidelberg. In lurid publications, he decried that Einstein was a "Jewish fraud" and that relativity "could have been predicted from the start—if race theory had been more widespread—since Einstein was a

Jew." Eventually, he became a leading member of what was called the Anti-relativity League, devoted to purging "Jewish physics" from Germany and establishing the purity of Aryan physics. Lenard was by no means alone within the physics world. He was joined by many in the German scientific establishment, including Nobel laureate Johannes Stark and Hans Geiger (inventor of the Geiger counter).

In August 1920, this virulent group of detractors booked Berlin's huge Philharmonic Hall strictly for the purpose of denouncing relativity theory. Remarkably, Einstein was in the audience. He braved a nonstop series of angry speakers who denounced him as a publicity hound, plagiarist, and charlatan to his face. The next month, there was yet another such confrontation, this time at a meeting of the Society of German Scientists in Bad Nauheim. Armed police were present to guard the hall's entrance and dampen any demonstration or violence. Einstein was jeered and hooted down when he tried to answer some of Lenard's inflammatory charges. News of this raucous exchange hit the papers in London, and the people of Britain became alarmed by rumors that Germany's great scientist was being hounded out of Germany. The German Foreign Office representative in London, to quell such rumors, said it would be catastrophic for German science if Einstein left, and that "we should not drive away such a man . . . whom we can use in effective cultural propaganda."

In April of 1921, with invitations pouring in from all corners of the world, Einstein decided to use his new celebrity to promote not only relativity but also his other causes, which now included peace and Zionism. He had finally rediscovered his Jewish roots. In long conversations with his friend Kurt Blumenfeld, he began to fully appreciate the deep suffering inflicted on the Jewish people throughout the centuries. Blumenfeld, he wrote, was responsible for "making me conscious of my Jewish soul." Chaim Weizmann, a leading Zionist, focused on the idea of using Einstein as a magnet to attract funds for Hebrew University in Jerusalem. The plan involved sending Einstein on a tour through the heartland of America.

As soon as Einstein's ship docked in New York Harbor, he was mobbed by reporters eager for a glimpse of him. Crowds lined the streets of New York to view his motorcade and cheered when he waved back from his open-topped limousine. "It's like the Barnum circus!" Elsa said, as someone threw a bouquet at her. Einstein mused, "The ladies of New York want to have a new style every year. This year the fashion is relativity." He added, "Do I have something of a charlatan or a hypnotist about me that draws people like a circus clown?"

As expected, Einstein aroused intense interest among the public and galvanized the Zionist cause. Well-wishers, curiosity seekers, and Jewish admirers packed every auditorium he spoke in. A mob of eight thousand squeezed into the Sixty-ninth Regiment Armory in Manhattan while three thousand had to be turned away, eagerly awaiting a glimpse of the genius. Einstein's reception at City College of New York was one of the highlights of the trip. Isidor Isaac Rabi, who would later win a Nobel Prize, took copious notes of Einstein's lecture and marveled that Einstein, unlike other physicists, possessed a crowd-pleasing charisma. (Even today, a picture of the entire student body of City College of New York crowding around Einstein hangs in the chairman's office at the school.)

After leaving New York, Einstein's trip through the United States was like a whistle-stop tour, passing through several major cities. In Cleveland, three thousand people mobbed him. He escaped "from possibly serious injury only by strenuous efforts by a squad of Jewish war veterans who fought the people off in their mad efforts to see him." In Washington, he met with President Warren G. Harding. Unfortunately, they could not communicate, since Einstein spoke no English and Harding did not speak German or French. (In all, Einstein's whirlwind tour netted almost a million dollars, $250,000 alone from a single dinner at the Waldorf Astoria Hotel speaking before eight hundred Jewish doctors.)

His travels in America not only introduced millions of Americans to the mystery of space and time, but also reaffirmed Einstein's deep and heartfelt commitment to the Jewish cause. Growing up in a comfortable, middle-class European family, he had no direct contact with the suffering of poor Jews from around the world. "It was the first time in my life that I saw Jews en masse," he noted. "Not until I was in America did I discover the Jewish people. I had seen many Jews, but neither in Berlin nor elsewhere in Germany had I encountered the Jewish people. The Jewish people I saw in America came from Russia, from Poland, or generally from eastern Europe."

After the United States, Einstein went to England, where he met the Archbishop of Canterbury. To the relief of the clergy, Einstein assured him that relativity theory would not undermine people's morale and belief in religion. He lunched at the Rothschilds and met the great classical physicist Lord Rayleigh, who said to Einstein, "If your theories are sound, I understand . . . that events, say, of the Norman Conquest have not yet occurred." When he was introduced to Lord Haldane and his daughter, she fainted at the sight of him. Later, Einstein paid homage to Isaac Newton by gazing at his tomb in England's most hallowed ground, Westminster Abbey, and laying a wreath. In March 1922, Einstein received an invitation to speak at the Collège de France, where he was mobbed by the Parisian press and followed by huge crowds. One journalist remarked, "He has become the great fashion. Academics, politicians, artists, policemen, cab drivers, and pickpockets know when Einstein lectures. *Tout* Paris knows everything and tells more than it knows about Einstein." Controversy surrounded the trip as some scientists, still nursing the wounds from World War I, boycotted his talk, using the excuse that they could not attend because Germany was not a member of the League of Nations. (In response, a Paris paper gibed, "If a German were to discover a cure for cancer or tuberculosis, would these thirty academics have to wait until Germany became a member of the League of Nations to use it?")

Einstein's return to Germany, however, was marred by the political instability of postwar Berlin. Ominously, it had become the season for political assassinations. In 1919, the socialist leaders Rosa Luxemburg and Karl Liebknecht had been killed. In April 1922, Walther Rathenau, a Jewish physicist and colleague of Einstein who had risen to become the German foreign minister, was assassinated by submachine guns as he rode in his car. A few days later, Maximilian Harden, another prominent Jew, was severely wounded in another assassination attempt.

A day of national mourning was declared, with theaters, schools, and universities closing to honor Rathenau. A million people stood silently near the Parliament building where the funeral services were being held. However, Philipp Lenard refused to cancel his classes at the Physics Institute in Heidelberg. (Previously, he had even advocated killing Rathenau. On the day of national mourning, a group of workers tried to persuade Lenard to cancel his classes, but were drenched with water thrown from the second floor of his building. The workers then broke into the institute and dragged Lenard out. They were about to throw him into the river when the police intervened.)

That year, a young German, Rudolph Leibus, was charged in Berlin with offering a reward for the murder of Einstein and other intellectuals, saying that "it was a patriotic duty to shoot these leaders of pacifist sentiment." He was found guilty by the courts but fined only sixteen dollars. (Einstein took these threats seriously, both from anti-Semites as well as deranged individuals. Once, a mentally unbalanced Russian immigrant, Eugenia Dickson, wrote a series of menacing letters to Einstein, raving that he was an imposter masquerading as the real Einstein, and stormed into Einstein's house trying to kill him. But before this crazed woman could attack Einstein, Elsa struggled with her at the door, managing to subdue her and call the police.)

Einstein, facing this dangerous tide of anti-Semitism, took the opportunity to launch another world tour, this time to the Orient. The philosopher and mathematician Bertrand Russell was on a speaking tour in Japan and was asked by his hosts to nominate some of the most illustrious people of the time to speak in Japan. He immediately nominated Lenin and Einstein. Since Lenin, of course, was unavailable, the invitation went to Einstein. He accepted it and began his odyssey in January 1923. "Life is like riding a bicycle. To keep your balance you must keep moving," he wrote.

While en route to Japan and China, Einstein received a message from Stockholm that many thought was long overdue. The telegram confirmed that he had won the Nobel Prize in physics. But he won the prize not for the relativity theory, his crowning achievement, but for the photoelectric effect. When Einstein finally delivered his Nobel Prize speech the next year, in typical fashion he shocked the audience by not speaking about the photoelectric effect at all, as everyone expected, but about relativity.

What took so long for Einstein, by far the most visible and respected figure in physics, to win the Nobel Prize? Ironically, he had been rejected eight times by the Nobel Prize Committee, from 1910 to 1921. During that period, numerous experiments had been conducted to verify the correctness of relativity. Sven Hedin, a member of the Nobel nominating committee, later confessed that the problem was Lenard, who had great influence over other judges, including Hedin. The Nobel Prize–winning physicist Robert Millikan also recalled that the Nobel nominating committee, split on the question of relativity, finally gave a committee member the task of evaluating the theory: "He spent all his time studying Einstein's theory of relativity. He couldn't understand it. Didn't dare to give the prize and run the risk of learning later that the theory of relativity is invalid."

As promised, Einstein sent the Nobel Prize money to Mileva as part of their divorce settlement ($32,000 in 1923 dollars). She would eventually use the money to purchase three apartment houses in Zurich.

By the 1920s and 1930s, Einstein had emerged as a giant on the world stage. Newspapers clamored for interviews, his face smiled from film newsreels, he was flooded with requests beseeching him to speak, and journalists would breathlessly print every trivial tidbit from his life. Einstein quipped that he was like King Midas, except everything he touched turned into a newspaper headline. New York University's Class of 1930, asked to name the world's most popular figure, chose Charles Lindbergh first, and Albert Einstein second, outranking all of Hollywood's movie stars. Everywhere Einstein went, his mere presence would spark huge crowds. For example, four thousand people started a near riot trying to crash a film explaining relativity at the American Museum of Natural History in New York. A group of industrialists even bankrolled the building of the Einstein Tower in Potsdam, Germany, a futuristic-looking solar observatory finished in 1924 that housed a telescope in a tower 54 feet high. Einstein was so much in demand from artists and photographers that wanted to capture the face of genius that he listed his job description as "artists' model."

This time, however, he did not make the mistake he made with Mileva, neglecting her while he was on world tours. To Elsa's delight, he took her along to greet celebrities, royalty, and the powerful. Elsa, in turn, adored her husband and gloried in his world fame. She was "gentle, warm, motherly, and proto-typically bourgeois, [and] loved to take care of her Albertle."

In 1930, Einstein made his second triumphant trip to the United States. On his visit to San Diego, the humorist Will Rogers noted about Einstein, "He ate with everybody, talked with everybody, posed for everybody that had any film left, attended every luncheon, every dinner, every movie opening, every marriage and two-thirds of the divorces. In fact, he made himself such a good fellow that nobody had the nerve to ask what his theory was." He visited the California Institute of Technology and the observatory at Mt. Wilson, meeting astronomer Edwin Hubble, who had verified some of Einstein's theories about the universe. He also visited Hollywood and received a glittering reception worthy of a superstar. In 1931, he and Elsa attended the world premier of Charlie Chaplin's film *City Lights.* The crowds strained to catch a fleeting glimpse of the world-famous scientist surrounded by Hollywood royalty. At the opening, as the audience wildly cheered Chaplin and Einstein, Chaplin remarked, "The people applaud me because everyone understands me, and they applaud you because no one understands you." Einstein, bewildered by the frenzy that celebrities can generate, asked what it all meant. Chaplin wisely replied, "Nothing." (When he visited New York's famed Riverside Church, he saw his face on a stained-glass window portraying the world's great philosophers, leaders, and scientists. He quipped, "I could have imagined they would make a Jewish saint out of me, but I never thought I would become a Protestant one!")

Einstein was also sought out for his thoughts on philosophy and religion. His meeting with a fellow Nobel laureate, Indian mystic Rabindranath Tagore, in 1930 attracted considerable press attention. They made quite a pair, with Einstein's flaming white hair and Tagore's equally imposing long white beard. One journalist remarked, "It was interesting to see them together— Tagore, the poet with the head of a thinker, and Einstein, the thinker with the head of a poet. It seemed to an observer as though two planets were engaged in a chat."

Ever since he read Kant as a child, Einstein became suspicious of traditional philosophy, which he often thought degenerated into pompous but ultimately simplistic hocus-pocus. He wrote,

"Is not all of philosophy as if written in honey? It looks wonderful when one contemplates it, but when one looks again it is all gone. Only mush remains." Tagore and Einstein clashed over the question of whether the world can exist independently of human existence. While Tagore held the mystical belief that human existence was essential to reality, Einstein replied, "The world, considered from the physical aspect, does exist independently of human consciousness." Although they disagreed on the question of physical reality, they found a bit more agreement on questions of religion and morality. In the area of ethics, Einstein believed that morality was defined by humanity, not by God. "Morality is of the highest importance—but for us, not God," Einstein observed. "I do not believe in the immorality of the individual, and I consider ethics to be an exclusively human concern with no superhuman authority behind it."

Although skeptical about traditional philosophy, he also had the deepest respect for the mysteries posed by religion, especially the nature of existence. He would write, "Science without religion is lame, religion without science is blind." He would also attribute this appreciation of mystery as the source of all science: "all the fine speculations in the realm of science spring from a deep religious feeling." Einstein wrote, "The most beautiful and deepest experience a man can have is the sense of the mysterious. It is the underlying principle of religion as well as of all serious endeavor in art and science." He concluded, "If something is in me which can be called religious, then it is the unbounded admiration for the structure of the world so far as science can reveal it." Perhaps his most elegant and explicit statement about religion was written in 1929: "I'm not an atheist and I don't think I can call myself a pantheist. We are in the position of a little child entering a huge library filled with books in many different languages. The child knows someone must have written those books. It does not know how. It does not understand the languages in which they are written. The child dimly suspects a mysterious order in the arrangement of the books but doesn't know what it is. That, it seems to me, is the attitude of even the most intelligent human being toward God. We see a universe marvelously arranged and obeying certain laws, but only dimly understand these laws. Our limited minds cannot grasp the mysterious force that moves the constellations. I am fascinated by Spinoza's pantheism, but admire even more his contributions to modern thought because he is the first philosopher to deal with the soul and body as one, not two separate things."

Einstein would often make a distinction between two types of Gods, which are often confused in discussions about religion. First, there is the personal God, the God that answers prayers, parts the waters, and performs miracles. This is the God of the Bible, the God of intervention. Then there is the God that Einstein believed in, the God of Spinoza, the God that created the simple and elegant laws that govern the universe.

Even in the midst of this media circus, Einstein miraculously never lost his focus and devoted his efforts to probing these laws of the universe. While on transatlantic ships or long train rides, he had the discipline to shut out distractions and concentrate on his work. And what intrigued Einstein during this period was the ability of his equations to solve the structure of the universe itself.

CHAPTER 6

The Big Bang and Black Holes

Did the universe have a beginning? Is the universe finite or infinite? Will it have an end? As he began to ask what his theory might say about the cosmos, Einstein, like Newton before him, encountered the same kinds of questions that had puzzled physicists centuries earlier.

In 1692, five years after Newton completed his masterpiece, *Philosophiae Naturalis Principia Mathematica*, he received a letter from a minister, Richard Bentley, that perplexed him. Bentley pointed out that if gravity was strictly attractive, and never repulsive, then any static collection of stars will necessarily collapse in on itself. This simple but potent observation was puzzling, as the universe seemed stable enough, yet his universal gravitation would, given enough time, collapse the entire universe! Bentley was isolating a key problem faced by any cosmology in which gravity was an attractive force: a finite universe must necessarily be unstable and dynamic.

After pondering this disturbing question, Newton wrote a letter back to Bentley, stating that the universe, to avoid this collapse, must therefore consist of an infinite, uniform collection of stars. If the universe were indeed infinite, then every star would be pulled evenly in all directions, and hence the universe could be stable even if gravity was strictly attractive. Newton wrote, "If the matter was evenly disposed throughout an infinite space, it could never convene into one mass . . .

and thus might the sun and fixed stars be formed."

But if one made that assumption, then there arose another, deeper problem, known as "Olbers' paradox." It asks, quite simply, why the night sky is black. If the universe is indeed infinite, static, and uniform, then everywhere we look, our eyes should see a star in the heavens. Thus, there should be an infinite amount of starlight hitting our eyes from all directions, and the night sky should be white, not black. So if the universe was uniform and finite, it would collapse, but if it were infinite, the sky should be on fire!

Over two hundred years later, Einstein faced the same problems, but in disguised form. In 1915, the universe was a comfortable place, thought to consist of a static, solitary galaxy, the Milky Way. This bright swath of light cutting across the night sky consisted of billions of stars. But when Einstein began to solve his equations, he found something disturbing and unexpected. He assumed that the universe was filled with a uniform gas, which approximated the stars and dust clouds. Much to his consternation, he found that his universe was dynamic, that it preferred to expand or contract and was never stable. In fact, he soon found himself in the quicksand of cosmological questions that have puzzled philosophers and physicists like Newton for ages. Finite universes are never stable under gravity.

Einstein, forced to confront a contracting or expanding dynamic universe like Newton, was still not ready to throw out the prevailing picture of a timeless, static universe. Einstein the revolutionary was still not revolutionary enough to accept that the universe was expanding or had a beginning. His solution was a rather feeble one. In 1917, he introduced what might be called a "fudge factor" into his equations, the "cosmological constant." This factor posited a repulsive antigravity that balanced the attractive force of gravity. The universe was made static by fiat.

To perform this sleight of hand, Einstein realized that general covariance, the main guiding mathematical principle behind general relativity, allows for two possible general covariant objects: the Ricci curvature (which forms the foundation of general relativity) and the volume of space-time. It was therefore possible to add a second term to his equations that was consistent with general covariance and proportional to the volume of the universe. In other words, the cosmological constant assigned an energy to empty space. This antigravity term, now called "dark energy," is the energy of the pure vacuum. It can push galaxies apart or bring them together. Einstein chose the value of the cosmological constant precisely to counteract the contraction caused by gravity, so the universe became static. He was unhappy with this, as it smacked of a mathematical swindle, but he had no choice if he wanted to preserve a static universe. (It would take another eighty years before astronomers finally found evidence for the cosmological constant, which is now believed to be the dominant source of energy in the universe.)

The puzzle deepened in the next few years as more solutions to Einstein's equations were discovered. In 1917, Willem de Sitter, a Dutch physicist, saw that it was possible to find a strange solution to Einstein's equations: a universe that was empty of all matter yet still expanded! All that was needed was the cosmological constant, the energy of the vacuum, to drive an expanding universe. This was unsettling to Einstein, who still believed, like Mach before him, that the nature of space-time should be determined by the matter content of the universe. Here was a universe that expanded without any matter whatsoever, needing only dark energy to propel itself forward.

The final radical steps were taken by Alexander Friedmann in 1922 and by a Belgian priest, Georges Lemaître, in 1927, who showed that an expanding universe emerges naturally from Einstein's equations. Friedmann obtained a solution of Einstein's equations beginning with a homogeneous, isotropic universe in which the radius expands or contracts. (Unfortunately,

Friedmann died in 1925 of typhoid fever in Leningrad before he could elaborate on his solution.) In the Friedmann-Lemaître picture, there are three possible solutions, depending on the density of the universe. If the density of the universe is larger than a certain critical value, then the expansion of the universe will eventually be reversed by gravity, and the universe will begin to contract. (The critical density is roughly ten hydrogen atoms per cubic yard.) In this universe, the overall curvature is positive (by analogy, a sphere has positive curvature). If the density of the universe is smaller than the critical value, then there is not enough gravity to reverse the expansion of the universe, so it expands indefinitely. (Eventually, the universe approaches near absolute zero in temperature as it expands toward what is called the "big freeze.") In this universe, the overall curvature is negative (by analogy, a saddle or a trumpet horn has negative curvature). Last, there is the possibility that the universe will be balanced right at the critical value (in which case it will still expand indefinitely). In this universe, the curvature is zero, so the universe is flat. Thus, the fate of the universe could be determined, in principle, by simply measuring its average density.

Progress in this direction was confusing, since now there were at least three cosmological models about how the universe should evolve (Einstein's, de Sitter's, and Friedmann-Lemaître's). The matter rested until 1929, when it was finally settled by the astronomer Edwin Hubble, whose results were to shake the foundations of astronomy. He first demolished the one-galaxy universe theory by demonstrating the presence of other galaxies far beyond the Milky Way. (The universe, far from being a comfortable collection of a hundred billion stars contained in a single galaxy, now contained billions of galaxies, each one containing billions of stars. In just one year, the universe suddenly exploded.) He found that there were potentially billions of other galaxies, and that the closest one was Andromeda, about two million light-years from Earth. (The word "galaxy," in fact, comes from the Greek word for "milk," since the Greeks thought that the Milky Way galaxy was milk spilled by the gods across the night sky.)

This shocking revelation alone would have guaranteed Hubble's fame among the giants of astronomy. But Hubble went further. In 1928, he made a fateful trip to Holland where he met de Sitter, who claimed that Einstein's general relativity predicted an expanding universe with a simple relationship between red shift and distance. The farther a galaxy was from Earth, the faster it would be moving away. (This red shift is slightly different from the red shift considered by Einstein back in 1915. This red shift is caused by galaxies receding from Earth in an expanding universe. If a yellow star, for example, moves away from us, the speed of the light beam remains constant but its wavelength gets "stretched," so that the color of the yellow star reddens. Similarly, if a yellow star approaches Earth, its wavelength is shrunken, squeezed like an accordion, and its color becomes bluish.)

When Hubble returned to the observatory at Mt. Wilson, he began a systematic determination of the red shift of these galaxies to see if this correlation held up. He knew that back in 1912, Vesto Melvin Slipher had shown that some distant nebulae were receding from Earth, creating a red shift. Hubble now systematically calculated the red shift coming from distant galaxies and discovered that these galaxies were receding from Earth—in other words, the universe was expanding at a fantastic rate. He then discovered that his data could fit the conjecture made by de Sitter. This is now called "Hubble's law": the faster a galaxy is receding from Earth, the farther it is (and vice versa).

Plotted on a curve, graphing distance versus velocity, Hubble found a near straight line, as predicted by general relativity, whose slope is now called "Hubble's constant." Hubble, in turn, was curious to know how his results would fit into Einstein's. (Unfortunately, Einstein's model had matter but no motion, and de Sitter's universe had motion but no matter. His results did seem to agree with the work of Friedmann and Lemaître, which possessed both matter and motion.) In 1930, Einstein made the pilgrimage to the Mt. Wilson observatory, where he met Hubble for the first time. (When the astronomers there proudly boasted that their mammoth 100-inch telescope, the biggest in the world at that time, could determine the structure of the universe, Elsa was not impressed. She said, "My husband does that on the back of an old envelope.") As Hubble explained the painstaking results he found from analyzing scores of galaxies, each one receding from the Milky Way, Einstein admitted that the cosmological constant was the greatest blunder of his life. The cosmological constant, introduced by Einstein to artificially create a static universe, was now dispensable. The universe did expand as he found a decade earlier.

Furthermore, Einstein's equations gave perhaps the simplest derivation of Hubble's law. Assume the universe is a balloon that is expanding, with the galaxies represented as tiny dots painted on the balloon. To an ant sitting on any one of these dots, it appears as if every other dot is moving away from it. Likewise, the farther a dot is away from the ant, the faster it is moving away, as in Hubble's law. Thus, Einstein's equations gave new insights into such ancient questions like, is there an end to the universe? If the universe ends with a wall, then we can ask the question, what lies beyond the wall? Columbus might have answered that question by considering the shape of the earth. In three dimensions, the earth is finite (being just a ball floating in space), but in two dimensions, it appears infinite (if one goes around and around its circumference) so anyone walking on the surface of the earth will never find the end. Thus, the earth is both finite and infinite at the same time, depending on the number of dimensions one measures. Likewise, one might state that the universe is infinite in three dimensions. There is no brick wall in space that represents the end of the universe; a rocket sent into space will never collide with some cosmic wall.

However, there is the possibility that the universe might be finite in four dimensions. (If it were a four-dimensional ball, or hypersphere, you might conceivably travel completely around the universe and come back to where you started. In this universe, the farthest object you can see with a telescope is the back of your head.)

If the universe is expanding at a certain rate, then one can reverse the expansion and calculate the rough time at which the expansion first originated. In other words, not only did the universe have a beginning, but also one could calculate its age. (In 2003, satellite data showed that the universe is 13.7 billion years old.) In 1931, Lemaître postulated a specific origin to the universe, a super-hot genesis. If one took Einstein's equations to their logical conclusion, they showed that there was a cataclysmic origin to the universe.

In 1949, cosmologist Fred Hoyle christened this the "big bang" theory during a discussion on BBC radio. Because he was pushing a rival theory, the legend got started that he coined the name "big bang" as an insult (although he later denied that story). However, it should be pointed out that the name is a complete misnomer. It was not big, and there was no bang. The universe started out as an infinitesimally small "singularity." And there was no bang or explosion in the conventional sense, since it was the expansion of space itself that pushed the stars apart.

Not only did Einstein's theory of general relativity introduce entirely unexpected concepts such as the expanding universe and the big bang, but also it introduced another concept that has intrigued astronomers ever since: black holes. In 1916, just one year after he published his theory of general relativity, Einstein was astonished to receive word that a physicist, Karl Schwarzschild, had solved his equations exactly for the case of a single pointlike star. Previously, Einstein had only used approximations to the equations of general relativity because they were so complex. Schwarzschild delighted Einstein by finding an exact solution, with no approximations whatsoever. Although Schwarzschild was director of the Astrophysical Observatory in Potsdam, he volunteered to serve Germany on the Russian front. Remarkably, as a soldier dodging shells bursting overhead, he managed against all odds to work on physics. Not only did he calculate the trajectory of artillery shells for the German army, he also calculated the most elegant, exact solution of Einstein's equations. Today, this is called the "Schwarzschild solution." (Unfortunately, he never lived long enough to enjoy the fame that his solution generated. One of the brightest stars emerging in this new field of relativity, Schwarzschild died at the age of forty-two, just a few months after his papers were published, from a rare skin disease he picked up while fighting on the Russian front, a waste for the world of science. Einstein delivered a moving eulogy for Schwarzschild, whose death only re-enforced Einstein's hatred of the senselessness of war.)

The Schwarzschild solution, which created quite a sensation in scientific circles, also had strange consequences. Schwarzschild found that extremely close to this pointlike star, gravity was so intense that even light itself could not escape, so the star became invisible! This was a sticky problem not only for Einstein's theory of gravity but also for the Newtonian theory. Back in 1783, John Michell, rector of Thornhill in England, posed the question whether a star could become so massive that even light could not escape. His calculations, using only Newtonian laws, could not be trusted because no one knew precisely what the speed of light was, but his conclusions were hard to dismiss. In principle, a star could become so massive that its light would orbit around it. Thirteen years later, in his famous book *Exposition du système du monde,* mathematician Pierre-Simon Laplace also asked whether these "dark stars" were possible (but probably found the speculation so wild that he deleted it from the third edition). Centuries later, the question of dark stars came up again, thanks to Schwarzschild. He found that there was a "magic circle" surrounding the star, now called the "event horizon," at which mind-bending distortions of space-time occur. Schwarzschild demonstrated that anyone unfortunate enough to fall past this event horizon would never return. (You would have to go faster than the speed of light to escape, which is impossible.) In fact, from inside the event horizon, nothing can escape, not even a light beam. Light emitted from this pointlike star would simply orbit around the star forever. From the outside, the star would appear shrouded in darkness.

One could use the Schwarzschild solution to calculate how much ordinary matter had to be compressed to reach this magic circle, called the "Schwarzschild radius," at which point the star would completely collapse. For the sun, the Schwarzschild radius was 3 kilometers, or less than 2 miles. For the earth, it was less than a centimeter. (Since this compression factor was beyond physical comprehension in the 1910s, physicists assumed that no one would ever encounter such a fantastic object.) But the more Einstein studied the properties of these stars, later christened "black holes" by physicist John Wheeler, the stranger they became. For example, if you fell into a black hole, it would only take a fraction of a second to fall through the event horizon. As you briefly sailed past it, you would see light orbiting the black hole that was captured perhaps aeons—perhaps billions of years—ago. The final millisecond would not be a very pleasant one. The gravitational forces would be so great that the atoms of your body would be crushed. Death would be inevitable and horrible. But observers watching this cosmic death take place from a safe distance would see an entirely different picture. The light emitted from your body would be stretched by gravity, so it would appear as if you were frozen in time. To the rest of the universe, you would still be hovering over the black hole, motionless.

These stars, in fact, were so fantastic that most physicists thought they could never be found in the universe. Eddington, for example, said, "There should be a law of Nature to prevent a star from behaving in this absurd way." In 1939, Einstein tried to show mathematically that such a black hole was impossible. He began by studying a star in formation, that is, a collection of particles circling around in space, gradually pulled in by their gravitational force. Einstein's calculation showed that this circling collection of particles would gradually collapse, but would only come within 1.5 times the Schwarzschild radius, and hence a black hole could never form.

Although this calculation seemed airtight, what Einstein apparently missed was the possibility of an implosion of matter in the star itself, created by the crushing effect of the gravitational force overwhelming all the nuclear forces in matter. This more detailed calculation was published in 1939 by J. Robert Oppenheimer and his student Hartland Snyder. Instead of assuming a collection of particles circling in space, they assumed a static star, large enough so that its massive gravity could overwhelm the quantum forces inside the star. For example, a neutron star consists of a large ball of neutrons about the size of Manhattan (20 miles across) making up a gigantic nucleus. What keeps this ball of neutrons from collapsing is the Fermi force, which prevents more than one particle with certain quantum numbers (e.g., spin) from being in the same state. If the gravitational force is large enough, then one can overcome the Fermi force and thereby squeeze the star to within the Schwarzschild radius, at which point nothing known to science can prevent a complete collapse. However, it would be another three decades or so before neutron stars were found and black holes were discovered, so most of the papers on the mind-bending properties of black holes were considered highly speculative.

Although Einstein was still rather skeptical about black holes, he was confident that one day yet another of his predictions would come true: the discovery of gravity waves. As we have seen, one of the triumphs of Maxwell's equations was the prediction that vibrating electric and magnetic fields would create a traveling wave that could be observed. Likewise, Einstein wondered if his equations allowed for gravity waves. In a Newtonian world, gravity waves cannot exist, since the "force" of gravity acts instantaneously throughout the universe, touching all objects simultaneously. But in general relativity, in some sense, gravity waves have to exist, as vibrations of the gravitational field cannot exceed the speed of light. Thus, a cataclysmic event, such as the collision of two black holes, will release a shock wave of gravity, a gravity wave, traveling at the speed of light.

As early as 1916, Einstein was able to show that with suitable approximations, his equations did yield wavelike motions of gravity. These waves spread across the fabric of space-time with the speed of light, as expected. In 1937, with his student Nathan Rosen, he was able to find an exact solution of his equations that gave gravity waves, with no approximations whatsoever. Gravity waves were now a firm prediction of general relativity. Einstein despaired, however, of ever being able to witness such an event. Calculations showed that it was far beyond the experimental capabilities of scientists at that time. (It would take almost eighty years, since Einstein first discovered gravity waves in his equations, for the Nobel Prize to be awarded to physicists who found the first indirect evidence for gravity waves. The first gravity waves may be directly detected perhaps ninety years after his first prediction. These gravity waves, in turn, may be the ultimate means by which to probe the big bang itself and find the unified field theory.)

In 1936, a Czech engineer, Rudi Mandl, approached Einstein with yet another idea concerning the strange properties of space and time, asking whether gravity from a nearby star could be used as a lens to magnify the light from distant stars, in the same way that glass lenses can be used to magnify light. Einstein had considered this possibility back in 1912, but, prodded by Mandl, calculated that the lens would create a ringlike pattern to an observer on Earth. For example, consider light from a faraway galaxy passing by a nearby galaxy. The gravity of the nearby galaxy might split the light in half, with each half going around the galaxy in opposite directions. When the light beams pass the nearby galaxy completely, they rejoin. From the earth, one would see these light beams as a ring of light, an optical illusion created by the bending of light around the nearby galaxy. However, Einstein concluded that "there is not much hope of observing this phenomenon directly." In fact, he wrote that this work "is of little value, but it makes the poor guy [Mandl] happy." Once again, Einstein was so far ahead of his time that it would take another sixty years before Einstein lenses and rings would be found and eventually become indispensable tools by which astronomers probe the cosmos.

As successful and far-reaching as general relativity was, it did not prepare Einstein in the mid-1920s for the fight of his life, to devise a unified field theory to unite the laws of physics while simultaneously doing battle with the "demon," the quantum theory.

PART III

THE UNFINISHED PICTURE

The Unified Field Theory

CHAPTER 7

Unification and the Quantum Challenge

I n 1905, almost as soon as Einstein worked out the special theory of relativity, he began to lose interest in it because he set his sights on bigger game: general relativity. In 1915, the pattern repeated itself. As soon as he finished formulating his theory of gravity, he began to shift his focus to an even more ambitious project: the unified field theory, which would unify his theory of gravity with Maxwell's theory of electromagnetism. It was supposed to be his masterpiece, as well as the summation of science's two-thousand-year investigation into the nature of gravity and light. It would give him the ability to "read the Mind of God."

Einstein was not the first to suggest a relationship between electromagnetism and gravity. Michael Faraday, working at the Royal Institution in London in the nineteenth century, performed some of the first experiments to probe the relationship between these two pervasive forces. He would, for example, drop magnets from the London Bridge and see if their rate of descent differed from that of ordinary rocks. If magnetism interacted with gravity, perhaps the magnetic field might act as a drag on gravity, making the magnets fall at a different rate. He would also drop pieces of metal from the top of a lecture room to a cushion on the floor, trying to see if the descent could induce an electric current in the metal. All his experiments produced negative results. However, he noted, "They do not shake my strong feeling of the existence of a relation between gravity and electricity, though they give no proof that such a relation exists." Furthermore, Riemann, who founded the theory of curved space in any dimension, believed strongly that both gravity and electromagnetism could be reduced to purely geometric arguments. Unfortunately, he did not have any physical picture or field equations, so his ideas went nowhere.

Einstein once summarized his attitude toward unification by comparing marble and wood. Marble, thought Einstein, described the beautiful world of geometry, in which surfaces warped smoothly and continuously. The universe of stars and galaxies played out their cosmic game on the beautiful marble of space-time. Wood, on the other hand, represented the chaotic world of matter, with its jungle of subatomic particles, its nonsensical rules for the quantum. This wood, like gnarled vines, grows in unpredictable and random ways. New particles being discovered in the atom made the theory of matter quite ugly. Einstein saw the defect in his equations. The fatal flaw was that wood determined the structure of the marble. The amount of bending of space-time was determined by the amount of wood at any point.

Thus, to Einstein, his strategy was clear: *to create a theory of pure marble,* to eliminate the wood by reformulating it solely in terms of marble. If the wood itself could be shown to be made of marble, then he would have a theory that was purely geometric. For example, a point particle is infinitely tiny, having no extension in space. In field theory, a point particle is represented by a "singularity," a point where the field strength goes to infinity. Einstein wanted to replace this singularity with a smooth deformation of space and time. Imagine, for example, a kink or knot in a rope. From a distance, the kink may look like a particle, but close-up the kink or knot is nothing but a wrinkle in the rope. Similarly, Einstein wanted to create a theory that was purely geometric and had no singularities whatsoever. Subatomic particles, like the electron, would emerge as kinks or as some kind of small wrinkle on the surface of space-time. The fundamental problem, however, was that he lacked a concrete symmetry and principle that could unify electromagnetism and gravity. As we saw earlier, the key to Einstein's thinking was unification through symmetry. With special relativity, he had the picture that guided him constantly, running next to a light beam. This picture revealed the fundamental contradiction between Newtonian mechanics and Maxwell's fields. From this, he was able to extract a principle, the constancy of the speed of light. Last, he was able to formulate the symmetry that unified space and time, the Lorentz transformations.

Similarly, with general relativity, he had a picture, that gravity was caused by the warping of space and time. This picture exposed the fundamental contradiction between Newton's gravity (where gravity traveled instantaneously) and relativity (where nothing can go faster than light). From the picture, he extracted a principle, the equivalence principle, that accelerating and gravitating frames obeyed the same laws of physics. Last, he was able to formulate the generalized symmetry that described accelerations and gravity, which was general covariance.

The problem facing Einstein now was truly daunting, because he was working at least fifty years ahead of his time. In the 1920s, when he began work on the unified field theory, the only established forces were the gravitational and electromagnetic forces. The nucleus of the atom had only been discovered in 1911 by Ernest Rutherford, and the force that held it together was still shrouded in mystery. But without an understanding of the nuclear forces, Einstein lacked a crucial part of the puzzle. Furthermore, no experiment or observation exposed a contradiction between gravity and electromagnetism that would be the hook Einstein could grab onto.

Hermann Weyl, a mathematician who was inspired by Einstein's search for a unified field theory, made the first serious attempt in 1918. At first, Einstein was very impressed. "It is a masterful symphony," he wrote. Weyl expanded Einstein's old theory of gravity by adding the Maxwell field directly into the equations. Then he demanded that the equations be covariant under even more symmetries than Einstein's original theory, including scale transformations (i.e., transformations that expand or contract all distances). However, Einstein soon found some strange anomalies in the theory. For example, if you traveled in a circle and came back to your original point, you would find that you were shorter but had the same shape. In other words, lengths were not preserved. (In Einstein's theory, lengths could also change, but they remained the same if you came back to where you started.) Time would also be shifted in a closed path, but this would violate our understanding of the physical world. For example, it meant that if vibrating atoms were moved around a complete circle, they would be vibrating at a different frequency when they came back. Although Weyl's theory seemed ingenious, it had to be abandoned because it did not fit the data. (In hindsight, we can see that the Weyl theory had too much symmetry. Scale invariance is apparently a symmetry that nature does not use to describe our visible universe.)

In 1923, Arthur Eddington also caught the bug. Inspired by Weyl's work, Eddington (and many others after him) tried his hand at a unified field theory. Like Einstein, he created a theory based on the Ricci curvature, but the concept of distance did not appear in the equations. In other words, it was impossible to define meters or seconds in his theory; the theory was "pre-geometrical." Only in the last step would distance finally appear as a consequence of his equations. Electromagnetism was supposed to emerge as a piece of the Ricci curvature. The physicist Wolfgang Pauli did not like this theory at all, stating that it had "no significance for physics." Einstein also panned it, thinking it had no physical content.

But what really rocked Einstein to the core was a paper that he saw in 1921, written by an obscure mathematician, Theodr Kaluza, from the University of Königsberg. Kaluza suggested that Einstein, who had pioneered the concept of the fourth dimension, add yet another dimension to his equations. Kaluza began by reformulating Einstein's own general relativity in *five dimensions* (four dimensions of space and one dimension of time). This takes no work at all, since Einstein's equations could easily be formulated in any dimension. Then, in a few lines, Kaluza showed that if the fifth dimension is separated from the other four, Einstein's equations emerged, along with Maxwell's equations! In other words, Maxwell's equations, the horrible set of eight partial differential equations memorized by every engineer and physicist, can be reduced to waves traveling on the fifth dimension. To put it another way, Maxwell's theory was already hidden inside Einstein's theory if relativity were extended to five dimensions.

Einstein was surprised by the sheer audacity and beauty of Kaluza's work. He wrote Kaluza, "The idea of achieving [unification] by means of a five-dimensional cylinder world never dawned on me. . . . At first glance, I liked your idea enormously." A few weeks later, after studying the theory, he wrote, "The formal unity of your theory is startling." In 1926, the mathematician Oskar Klein generalized Kaluza's work and speculated that the fifth dimension was unobservable because it was small and possibly linked to the quantum theory. Kaluza and Klein were thus proposing an entirely different approach to unification. To them, electromagnetism was nothing but vibrations rippling along the surface of a small fifth dimension.

For example, if we think of fish living in a shallow pond, swimming just below the lily pads, the fish might conclude that their universe was two-dimensional. They can move forward and backward, left and right, but the concept of "up" into the third dimension would be alien to them. If their universe was two-dimensional, then how might they become aware of a mysterious third dimension? Imagine that it rains one day. Tiny ripples in the third dimension move along the surface of the pond, and they are clearly visible to the fish. As these ripples move along the surface, the fish might conclude that there was a mysterious force that could illuminate their universe. Similarly, in this picture, we are the fish. We conduct our affairs in three spatial dimensions, unaware that there could be higher dimensions existing just beyond our senses. The only direct contact that we might have with the unseen fifth dimension is light, now viewed as ripples traveling along the fifth dimension.

There was a reason why the Kaluza-Klein theory worked so well. Recall that *unification through symmetry* was one of Einstein's great strategies that led to relativity. In the Kaluza-Klein theory, electromagnetism and gravity were united because of a new symmetry, five-dimensional general covariance. Although this picture was immediately appealing, unifying gravity and electromagnetism by introducing another dimension, there was still the nagging question, where was this fifth dimension? No experiment has ever, even to this day, picked up evidence of any higher dimension of space beyond length, width, and height. If these higher dimensions exist, then they must be extremely small, much smaller than an atom. For example, we know that if we release chlorine gas into a room, its atoms can slowly permeate all the nooks and crannies of any room without disappearing into some mysterious extra dimension. We know, therefore, that any hidden dimension must be smaller than any atom. In this new theory, if one makes the fifth dimension smaller than an atom, then it is consistent with all laboratory measurements, which have never detected the presence of the fifth dimension. Kaluza and Klein assumed that the fifth dimension was "curled up" into a small ball, too small to be experimentally observed.

Although the Kaluza-Klein theory was a fresh, intriguing approach to unifying electromagnetism with gravity, Einstein eventually had doubts. The thought that the fifth dimension might not exist, that it might be a mathematical fiction or mirage, bothered him. Also, he had problems finding subatomic particles in the Kaluza-Klein theory. His goal was to derive the electron from his gravitational field equations, and try as he could, he could find no such solution. (In hindsight, this was a tremendous missed opportunity for physics. If physicists had taken the Kaluza-Klein theory more seriously, they might have added more dimensions beyond five. As we increase the number of dimensions, Maxwell's field increases in number into what are called "Yang-Mills fields." Klein actually discovered the Yang-Mills fields in the late 1930s, but his work was forgotten because of the chaos of World War II. It would take almost two decades before they were rediscovered, in the mid-1950s. These Yang-Mills fields now form the foundation of the current theory of the nuclear force. Almost all of subatomic physics is formulated in terms of them. After another twenty years, the Kaluza-Klein theory would be resurrected in the form of a new theory, string theory, now considered the leading candidate for a unified field theory.)

Einstein hedged his bets. If the Kaluza-Klein theory failed, then he would have to explore a different avenue toward the unified field theory. His choice was to investigate geometries beyond Riemannian geometry. He consulted many mathematicians, and it became quickly obvious that this was a totally open field. In fact, at Einstein's urging, many mathematicians began to look into "post-Riemannian" geometries, or the "theory of connections," to help him explore new possible universes. New geometries involving "torsion" and "twisted spaces" were soon developed as a consequence. (These abstract spaces would have no application to physics for another seventy years, until the arrival of superstring theory.)

Working on post-Riemannian geometries was a nightmare, however. Einstein had no guiding physical principle to help him through the thicket of abstract equations. Previously, he used the equivalence principle and general covariance as compasses. Both were firmly rooted in experimental data. He had also relied on physical pictures to show him the way. With the unified field theory, however, he had no guiding physical principle or picture.

So curious was the world about Einstein's work that a progress report he gave on the unified field theory to the Prussian Academy was reported to the *New York Times*, which even published parts of Einstein's paper. Soon, there were hundreds of reporters swarming outside his home, hoping for a glimpse of him. Eddington wrote, "You may be amused to hear that one of our great department stores in London (Selfridges) has posted on its window your paper (the six pages pasted up side by side) so that passers-by can read it all through. Large crowds gather around to read it." Einstein, however, would have traded all the adulation and praise in the world for a simple physical picture to guide his path.

Gradually, other physicists began to hint that Einstein was on the wrong track and that his physical intuition was failing him. One critic was his friend and colleague Wolfgang Pauli, one of the early pioneers of the quantum theory, who was famous in scientific circles for his unsparing wit. He once said of a misguided physics paper, "It is not even wrong." To a colleague whose paper he had reviewed, he said, "I do not mind if you think slowly, but I do object when you publish more quickly than you think." After he heard a confused, incoherent seminar, he would say, "What you said was so confusing that one cannot tell whether it was nonsense or not." When fellow physicists complained that Pauli was too critical, he would reply, "Some people have very sensitive corns, and the only way to live with them is to step on these corns until they are used to it." His impression of the unified field theory was reflected by his famous comment that what God has torn asunder, let no man put together. (Ironically, later Pauli would also catch the bug and propose his own version of the unified field theory.)

Pauli's view would have been endorsed by many of his fellow physicists, who grew increasingly preoccupied with the quantum theory, the other great theory of the twentieth century. The quantum theory stands as one of the most successful physical theories of all time. It has had unparalleled success explaining the mysterious world of the atom, and by doing so has unleashed the power of lasers, modern electronics, computers, and nanotechnology. Ironically, however, the quantum theory is based on a foundation of sand. In the atomic world, electrons seemingly appear in two places at the same time, jump between orbits without warning, and disappear into the ghostly world between existence and nonexistence. As Einstein remarked as early as 1912, "The more success the quantum theory has, the sillier it looks."

Some of the bizarre features of the quantum world were made apparent in 1924, when Einstein received a curious letter from an obscure Indian physicist, Satyendra Nath Bose, whose papers on statistical physics were so strange that they were flatly rejected for publication. Bose was proposing an extension of Einstein's earlier work on statistical mechanics, searching for a fully quantum mechanical treatment of a gas, treating the atoms as quantum objects. Just as Einstein had extended Planck's work to a theory of light, Bose was hinting that one could extend Einstein's work into a fully quantum theory of atoms in a gas. Einstein, a master of the subject, found that though Bose had made a number of mistakes, making assumptions that could not be justified, his final answer appeared to be correct. Einstein was not only intrigued by the paper, he translated it into German and submitted it for publication.

He then extended Bose's work and wrote a paper of his own,

applying the result to extremely cold matter that hovers just above the temperature of absolute zero. Bose and Einstein found a curious fact about the quantum world: all atoms are indistinguishable; that is, you cannot put a label on each atom, as Boltzmann and Maxwell had thought. While rocks and trees and other ordinary matter can be labeled and given names, in the quantum world all atoms of hydrogen are identical in any experiment; there are no green or blue or yellow hydrogen atoms. Einstein then found that if a collection of atoms were supercooled to near absolute zero, where all atomic movement almost ceases, all the atoms would fall down to the lowest energy state, creating a single "superatom." These atoms would condense into the same quantum state, behaving essentially like a single gigantic atom. He was proposing an entirely new state of matter, never seen before on Earth. However, before the atoms could tumble down to the lowest energy state, the temperatures would have to be fantastically small, much too small to be experimentally observed, about a millionth of a degree above absolute zero. (At these extremely low temperatures, the atoms vibrate in lockstep, and subtle quantum effects only seen at the level of individual atoms now become distributed throughout the entire condensate. Like the spectators at a football game who form "human waves" that sweep across the stands as they stand up and down in unison, the atoms in a "Bose-Einstein condensate" act as if everything is vibrating in unison.) But Einstein despaired of ever observing this Bose-Einstein condensation in his lifetime, since the technology of the 1920s did not permit experiments near temperatures of absolute zero. (In fact, Einstein was so ahead of his time that it would be about seventy years before that prediction could be tested.)

In addition to Bose-Einstein condensation, Einstein was interested in whether his principle of duality could be applied to matter as well as light. In his 1909 lecture, Einstein had showed that there was a dual nature to light, that it can simultaneously have both particle and wavelike properties. Although it was a heretical idea, it was supported fully by experimental results. Inspired by the duality program initiated by Einstein, a young graduate student, Prince Louis de Broglie, then speculated in 1923 that even matter itself can have both particle and wavelike properties. This was a bold, revolutionary concept, since it was a deep-seated prejudice that matter consisted of particles. Stimulated by Einstein's work on duality, de Broglie could explain away some of the mysteries of the atom by introducing the concept that matter had wavelike properties.

Einstein liked the audacity of de Broglie's "matter waves" and promoted his theory. (De Broglie would eventually be awarded the Nobel Prize for this seminal idea.) But if matter had wavelike properties, then what was the equation that the waves obeyed? Classical physicists had plenty of experience writing down the equations of ocean waves and sound waves, so an Austrian physicist, Erwin Schrödinger, was inspired to write down the equation of these matter waves. While Schrödinger, a well-known ladies' man, was staying with one of his innumerable girlfriends in the Villa Herwig in Arosa during the Christmas holidays in 1925, he managed to divert himself long enough to formulate an equation that would soon be known as one of the most celebrated in all of quantum physics, the Schrödinger wave equation. Schrödinger's biographer, Walter Moore, wrote, "Like the dark lady who inspired Shakespeare's sonnets, the lady of Arosa may remain forever mysterious." (Unfortunately, because Schrödinger had so many girlfriends and lovers in his life, as well as illegitimate children, it is impossible to determine precisely who served as the muse for this historic equation.) Over the next several months, in a remarkable series of papers, Schrödinger showed that the mysterious rules found by Niels Bohr for the hydrogen atom were simple consequences of his equation. For the first time, physicists had a detailed picture of the interior of the atom, by which one could, in principle, calculate the properties of more complex atoms, even molecules. Within months, the new quantum theory became a steamroller, obliterating many of the most puzzling questions about the atomic world, answering the greatest mysteries that had stumped scientists since the Greeks. The dance of electrons as they moved between orbits, releasing pulses of light or binding molecules together, suddenly became calculable, a matter of solving standard partial differential equations. One young brash quantum physicist, Paul Adrian Maurice Dirac, even boasted that all of chemistry could be explained as solutions of Schrödinger's equation, reducing chemistry to applied physics.

Thus Einstein, who was the father of the "old quantum theory" of the photon, became the godfather of the "new quantum theory," based on these Schrödinger waves. (Today, when high school chemistry students have to memorize the funny football-shaped "orbitals" that surround the nucleus, with strange labels and "quantum numbers," they are actually memorizing the solutions to the Schrödinger wave equation.) The breakthroughs in quantum physics now accelerated enormously. Realizing that the Schrödinger equation did not incorporate relativity, just two years later Dirac generalized the Schrödinger equation to a fully relativistic theory of electrons, and once again the world of physics was dazzled. While Schrödinger's celebrated equation was nonrelativistic and only applied to electrons moving at slow velocities compared to light, Dirac's electrons obeyed the full Einstein symmetry. Furthermore, Dirac's equation could automatically explain some obscure properties of the electron, including something called "spin." It was known from earlier experiments by Otto Stern and Walter Gerlach that the electron acted like a spinning top in a magnetic field, with angular momentum given by ½ (in units of Planck's constant). The Dirac electron yielded precisely the spin ½ given by the Stern-Gerlach experiment. (The Maxwell field, representing the photon, has spin 1, and Einstein's gravity waves have spin 2. With Dirac's work, it became clear that the spin of a subatomic particle would be one of its important properties.)

Then Dirac went one step further. By looking at the energy of these electrons, he found that Einstein had overlooked a solution to his own equations. Usually, when taking the square root of a number, we introduce both positive and negative solutions. For example, the square root of 4 can be either plus 2 or minus 2. Because Einstein ignored a square root in his equations, his famous equation $E = mc^2$ was not quite correct. The correct equation was $E = \pm mc^2$. This extra minus sign, argued Dirac, made possible a new kind of mirror universe, one in which particles could exist with a new form of "antimatter."

(Strangely, just a few years earlier in 1925, Einstein himself had entertained the idea of antimatter when he showed that by reversing the sign of the electron charge in a relativistic equation, one can get identical equations if one also reverses the orientation of space. He showed that for every particle of a certain mass, there must exist another particle with opposite charge but identical mass. Relativity theory not only gave us the fourth dimension, it was now giving us a parallel world of antimatter. However, Einstein, never one to quibble over priority, graciously never challenged Dirac.)

At first, the radical ideas of Dirac met with fierce skepticism. The idea of an entire universe of mirror particles that arose from $E = \pm mc^2$ seemed like an outlandish idea. Quantum physicist Werner Heisenberg (who with Niels Bohr had independently found a formulation of the quantum theory equivalent to Schrödinger's) wrote, "The saddest chapter of modern physics is and remains the Dirac theory. . . . I regard the Dirac theory . . . as learned trash which no one can take seriously." However, physicists had to swallow their pride when the antielectron, or positron, was finally discovered in 1932, for which Dirac later received the Nobel Prize. Heisenberg finally admitted, "I think that this discovery of anti-matter was perhaps the biggest jump of all the big jumps in our century." Once again, the theory of relativity yielded unexpected riches, this time giving us an entirely new universe made of antimatter.

(It seems strange that Schrödinger and Dirac, who developed the two most important wave functions in the quantum theory, were such polar opposites in their personalities. While Schrödinger was always accompanied by some lady friend, Dirac was painfully shy with women and was a man of remarkably few words. After Dirac's death, the British, honoring his contributions to the world of physics, had the relativistic Dirac equation engraved into stone in Westminster Abbey, not far from Newton's grave.)

Soon, physicists at every institute on this planet struggled to learn the strange, beautiful properties of the Schrödinger and Dirac equations. However, for all their undeniable successes, quantum physicists still had to grapple with a troubling philosophical question: if matter is a wave, then precisely *what is waving*? This is the same question that had haunted the wave theory of light, which gave birth to the incorrect theory of the aether. A Schrödinger wave is like an ocean wave and eventually spreads out if left by itself. With enough time, the wave function eventually dissipates over the entire universe. But this violated everything that physicists knew about electrons. Subatomic particles were believed to be pointlike objects that made definite, jetlike streaks which could be photographed on film. Thus, although these quantum waves had near miraculous success in describing the hydrogen atom, it did not seem possible that the Schrödinger wave could describe an electron moving in free space. In fact, if the Schrödinger wave really represented an electron, it would slowly dissipate and the universe would dissolve.

Something was terribly wrong. Finally, Einstein's lifelong friend Max Born proposed one of the most controversial solutions to this puzzle. In 1926, Born took the decisive step, claiming that the Schrödinger wave did not describe the electron at all, but only the *probability* of finding the electron. He declared that "the motion of particles follows probability laws, but probability itself propagates in conformity with the laws of causality." In this new picture, matter indeed consisted of particles, not waves. The markings captured on photographic plates are the tracks left by pointlike particles, not waves. But the chance of finding the particle at any given point was given by a wave. (More precisely, the absolute square of the Schrödinger wave represents the probability of finding the particle at a specific point in space and time.) Thus, it did not matter if the Schrödinger wave spread out over time. It simply meant that if you left an electron by itself, over time it would wander around and you would not know precisely where it was. All the paradoxes were now solved: the Schrödinger wave was not the particle itself, but represented the chance of finding it.

Werner Heisenberg took this one step further. He had agonized endlessly with Bohr over the puzzles of probability infesting this new theory, often getting into heated arguments with his older colleague. One day, after a frustrating night of grappling with the question of probabilities, he took a long stroll down Faelled Park, behind his university, constantly asking himself how it was possible that one could not know the precise location of an electron. How can the location of an electron be uncertain, as claimed by Born, if you can simply measure where it is?

Then, it suddenly hit him. Everything became clear. In order to know where an electron was, you had to look at it. This meant shining a light beam at it. But the photons in the light beam would collide with the electron, making its position uncertain. In other words, the act of observation necessarily introduced uncertainty. He reformulated this question into a new principle of physics, the uncertainty principle, which states that *one cannot determine both the location and the velocity of a particle at the same time.* (More precisely, the product of the uncertainty in position and momentum must be greater than or equal to Planck's constant divided by 4π). This was not just a by-product of the crudeness of our instruments; it was a fundamental law of nature. Even God could not know both the precise position and momentum of an electron.

This was the decisive moment when the quantum theory plunged into deep, totally uncharted waters. Up to then, one could argue that quantum phenomena were statistical, representing the average motions of trillions of electrons. Now, even the motions of a single electron could not be definitively determined. Einstein was horrified. He almost felt betrayed, knowing that his good friend Max Born was abandoning determinism, one of the most cherished ideas in all of classical physics. Determinism states, in essence, that you can determine the future if you know everything about the present. For example, Newton's great contribution to physics was that he could predict the motion of comets, moons, and planets via his laws of motion once he knew the present state of the solar system. For centuries, physicists had marveled at the precision of Newton's laws, that they could predict the position of celestial bodies, in principle, millions of years into the future. In fact, up to that time, all of science was based on determinism; that is, a scientist can predict the outcome of an experiment if the scientist knows the position and velocities of all particles. Followers of Newton summarized this belief by comparing the universe to a gigantic clock. God wound up this clock at the beginning of time and it has been steadily ticking ever since according to Newton's laws of motion. If you knew the position and velocity of every atom in the universe, then you can, via Newton's laws of motion, calculate the subsequent evolution of the universe with infinite precision. However, the uncertainty principle negated all of this, stating that it is impossible to predict the future state of the universe. Given a uranium atom, for example, one could never calculate when it will decay, only the likelihood of its doing so. In fact, even God or a deity did not know when the uranium atom would decay.

In December 1926, responding to Born's paper, Einstein wrote, "Quantum mechanics calls for a great deal of respect. But some inner voice tells me that this is not the true Jacob. The theory offers a lot, but it hardly brings us any closer to the Old Man's secret. For my part, at least I am convinced that He doesn't throw dice." When commenting on Heisenberg's theory, Einstein remarked, "Heisenberg has laid a big quantum egg. In Göttingen they believe in it (I don't)." Schrödinger himself disliked this idea intensely. He once said that if his equation represented only probabilities, then he regretted having anything to do with it. Einstein chimed in that he would have become a "cobbler or employee in a gaming house," if he had known that the quantum revolution he helped to initiate would introduce chance into physics.

Physicists were beginning to divide into two camps. Einstein led one camp, which still clung to a belief in determinism, an idea that dated back to Newton himself and had guided physicists for centuries. Schrödinger and de Broglie were allies. The other, much larger camp was led by Niels Bohr, who believed in uncertainty and championed a new version of causality, based on averages and probabilities.

Bohr and Einstein, in some sense, were polar opposites in other ways. While Einstein as a child shunned sports and was glued to books on geometry and philosophy, Bohr was renowned throughout Denmark as a soccer star. Whereas Einstein spoke forcefully and dynamically, wrote almost lyrically, and could exchange banter with journalists as well as royalty, Bohr was stiff, had a horrible mumble, was often inarticulate and inaudible, and would often repeat a single word endlessly when engrossed in thought. While Einstein could effortlessly write elegant and beautiful prose, Bohr was paralyzed when he had to write a paper. As a high school student, he would dictate all his papers to his mother. After he married, he would dictate them to his wife (even interrupting his honeymoon to dictate one long and important paper). He would sometimes involve his entire laboratory in rewriting his papers, once over a hundred times, completely disrupting the work. (Wolfgang Pauli, once asked to visit Bohr in Copenhagen, replied, "If the last proof is sent away, then I will come.") Both were, however, obsessed with their first love, physics. Bohr, in fact, would scribble equations on the goal post of a soccer game if he had an inspiration. Both would also sharpen their thoughts by using others as sounding boards for their ideas. (Strangely, Bohr could only function if he had assistants around him to bounce off ideas. Without an assistant whose ear he could borrow, he was helpless.)

The showdown finally came at the Sixth Solvay Conference in Brussels in 1930. What was at stake was nothing less than the nature of reality itself. Einstein hammered incessantly at Bohr, who reeled under the constant attacks but managed to ably defend his positions. Finally, Einstein presented an elegant "thought experiment" which, he thought, would demolish the "demon," the uncertainty principle: Imagine a box containing radiation. There is a hole in the box with a shutter. When the shutter is opened briefly, it can release a single photon from the box. Thus, we can measure with great certainty the precise time at which the photon was emitted. Much later, the box can be weighed. Because of the release of the photon, the box weighs less. Because of the equivalence of matter and energy, we can now tell how much total energy the box contains, also to great accuracy. Thus, we now know both the total energy and the time of opening of the shutter to arbitrary accuracy, without any uncertainty, and hence the uncertainty principle is wrong. Einstein thought he had finally found the tool to demolish the new quantum theory.

Paul Ehrenfest, one of the participants to this conference and a witness to this fierce battle, would write, "To Bohr, this was a heavy blow. At the moment he saw no solution. He was extremely unhappy all through the evening, walked from one person to another, trying to persuade them all that this could not be true, because if E was right this would mean the end of physics. But he could think of no refutation. I will never forget the sight of the two opponents leaving the university club. Einstein, a majestic figure, walking calmly with a faint ironical smile, and Bohr trotting along by his side, extremely upset." When he talked to Ehrenfest later that evening, all Bohr could mumble was one word, over and over again: "Einstein . . . Einstein . . . Einstein." But after an intense, sleepless night, Bohr finally found the defect in Einstein's argument, and he used Einstein's own theory of relativity to defeat him. Bohr noted that because the box weighed less than before, it would rise slightly in the earth's gravity. But according to general relativity, time speeds up as gravity gets weaker (so that time beats faster on the moon, for example). Thus, any minuscule uncertainty in measuring the time of the shutter would be translated into an uncertainty in measuring the position of the box. You cannot, therefore, measure the position of the box with absolute certainty. Furthermore, any uncertainty in the weight of the box will be reflected in an uncertainty in its energy and also its momentum, and hence you cannot know the momentum of the box with absolute certainty. When everything is put together, the two uncertainties identified by Bohr, the uncertainty in position and uncertainty in momentum, agree precisely with the uncertainty principle. Bohr had successfully defended the quantum theory. When Einstein complained that "God does not play dice with the world," Bohr reportedly fired back, "Stop telling God what to do."

Ultimately, Einstein had to admit that Bohr had successfully refuted his arguments. Einstein would write, "I am convinced that this theory undoubtedly contains a piece of definitive truth." Commenting on the historic Bohr-Einstein debate, John Wheeler said it was "the greatest debate in intellectual history that I know about. In thirty years, I never heard of a debate between two greater men over a longer period of time on a deeper issue with deeper consequences for understanding this strange world of ours."

Schrödinger, who also hated this new interpretation of his equations, proposed his celebrated problem of the cat to poke holes into the uncertainty principle. Schrödinger wrote about quantum mechanics: "I don't like it, and I'm sorry I had anything to do with it." The most ridiculous problem, he wrote, was that of a cat sealed in a box, inside which there is a bottle of hydrocyanic acid, a poisonous gas, connected to a hammer, triggered by a Geiger counter that is connected to a piece of radioactive substance. There is no question that radioactive decay is a quantum effect. If the uranium does not decay, then the cat is alive. But if an atom decays, it will set off the counter, trigger the hammer, break the glass, and kill the cat. But according to the quantum theory, we cannot predict when the uranium atom will decay. In principle, it may exist in both states simultaneously, both intact and decayed. But if the uranium atom can exist simultaneously in both states, then it means that the cat must also exist in both states. So the question is, is the cat dead or alive?

Normally, this is a silly question. Even if we cannot open the box, common sense tells us that the cat is either dead or alive. One cannot be both dead and alive simultaneously; this would violate everything we know about the universe and physical reality. However, the quantum theory gives us a strange answer. The final answer is, we don't really know. Before you open the box, the cat is represented by a wave, and waves can add, like numbers. We have to add the wave function of a dead cat to that of a live cat. *Thus, the cat is neither dead nor alive before you open the box.* Sealed inside the box, all you can say is that there are waves that represent the cat being both dead and alive at the same time.

Once we finally open the box, we can make a measurement and see for ourselves if the cat is dead or alive. The measurement process, by an outside observer, allows us to "collapse" the wave function and determine the precise state of the cat. Then we know if the cat is dead or alive. The key is the measurement process by an outside observer; by shining a light inside the box, the wave function has collapsed and the object suddenly assumes a definitive state.

In other words, the process of observation determines the final state of an object. The weakness of Bohr's Copenhagen interpretation lies in the question, do objects really exist before you make a measurement? To Einstein and Schrödinger, all this seemed preposterous. For the rest of his life, Einstein would grapple with these deep philosophical questions (which even today are still the subject of intense debate).

Several upsetting aspects of this puzzle shook Einstein to the core. First, before a measurement is made, we exist as the sum of all possible universes. We cannot say for certain if we are dead or alive, or whether dinosaurs are still alive, or whether the earth was destroyed billions of years ago. All events, before a measurement is made, are possible. Second, it would seem that the process of observation creates reality! Thus, we have a new twist to the old philosophical puzzle of whether a tree really falls in the forest if no one hears it. A Newtonian would argue that the tree can fall, independent of observation. But someone from the Copenhagen school would say that the tree can exist in all possible states (fallen, upright, sapling, mature, burnt, rotten, etc.) until it is observed, at which point it suddenly springs into existence. Thus the quantum theory adds a totally unexpected interpretation: observing the tree *determines* the state of the tree, that is, whether it fell or not.

Einstein, from his days at the patent office, always had an uncanny knack for isolating the essence of any problem. He would, therefore, ask visitors to his home the following question: "Does the moon exist because a mouse looks at it?" If the Copenhagen school is correct, then yes, in some sense the moon springs into existence when a mouse observes it, and the moon's wave function collapses. Over the decades, a number of "solutions" have been offered to the cat problem, none of them totally satisfactory. Although almost no one challenges the validity of quantum mechanics itself, these questions still remain as some of the greatest philosophical challenges in all of physics.

"I have thought a hundred times as much about the quantum problems as I have about general relativity theory," wrote Einstein about how he endlessly grappled with the foundations of the quantum theory. After much deep thought, Einstein fired back with what he thought was the definitive critique of the quantum theory. In 1933, with his students Boris Podolsky and Nathan Rosen, he proposed a novel experiment that even today is causing headaches to many quantum physicists as well as philosophers. The "EPR experiment" may not have demolished quantum theory, as Einstein had hoped it would, but it succeeded in proving that the quantum theory, which was already pretty bizarre, gets weirder and weirder. Suppose that an atom emits two electrons in opposite directions. Each electron is spinning like a top, pointing either up or down. Suppose further that they are spinning in opposite ways, so the total spin is zero, although you don't know which way they are spinning. For example, one electron may be spinning up, while the other is spinning down. If you wait long enough, these electrons could be separated by billions of miles. Before any measurement is made, you don't know the spins of the electrons.

Now suppose that you finally measure the spin of one electron. It is, for example, found to be spinning up. Then instantly, you know the spin of the other electron, although it is many light-years away—since its spin is the opposite of its partner, it must be spinning down. This means that a measurement in one part of the universe instantly determined the state of an electron on the other side of the universe, seemingly in violation of special relativity. Einstein called this "spooky action-at-a-distance." The philosophical implications of this are rather startling. It means that some atoms in our body may be connected with an invisible web to atoms on the other side of the universe, such that motions in our body can instantly affect the state of atoms billions of light-years away, in seeming violation of special relativity. Einstein disliked this idea, because it meant that *the universe was nonlocal*; that is, events here on Earth instantly affect events on the other side of the universe, traveling faster than light.

On hearing of this new objection to quantum mechanics, Schrödinger wrote to Einstein, "I was very happy that in that paper . . . you have evidently caught dogmatic quantum mechanics by the coat-tails." Hearing of the latest Einstein paper, Bohr's colleague Leon Rosenfeld wrote, "We dropped everything; we had to clear up such a misunderstanding at once. Bohr, in great excitement, instantly began dictating the draft of a rejoinder."

The Copenhagen school withstood the challenge, but at a price: Bohr had to concede to Einstein that the quantum universe was indeed nonlocal (i.e., events in one part of the universe can instantly affect another part of the universe). Everything in the universe is somehow meshed together in a cosmic "entanglement." So the EPR experiment did not disprove quantum mechanics; it only revealed how crazy it really is. (Over the years, this experiment has been misunderstood, with scores of speculations that one could build EPR faster-than-light radio, or that we can send signals back in time, or that we can use this effect for telepathy.)

The EPR experiment did not negate relativity, however. In this sense, Einstein had the last laugh. No useful information can be transmitted faster than light via the EPR experiment. For example, you cannot send Morse code faster than light via the

EPR apparatus. Physicist John Bell used this example to explain the problem. He described a mathematician called Bertlmann who always wore a pink sock and a green sock. If you knew that one foot had the green sock, you knew immediately that the other sock was pink. Yet no signal went from one foot to the other. In other words, knowing something is entirely different from sending that knowledge. There is a world of difference between the possession of information and its transmission.

By the late 1920s, there were now two towering branches of physics: relativity and the quantum theory. The sum total of all human knowledge about the physical universe could be summarized by these two theories. One theory, relativity, gave us a theory of the very large, a theory of the big bang and black holes. The other theory, the quantum theory, gave us a theory of the very small, the bizarre world of the atom. Although the quantum theory was based on counterintuitive ideas, no one could dispute its stunning experimental successes. Nobel Prizes were practically flying off the wall for young physicists willing to apply the quantum theory. Einstein was too seasoned a physicist to ignore the breakthroughs being made almost daily in the quantum theory. He did not dispute the experimental successes of it. Quantum mechanics was the "most successful physical theory of our period," he would admit. Neither did Einstein impede the development of quantum mechanics, as a lesser physicist might have. (In 1929, Einstein recommended that Schrödinger and Heisenberg share in the Nobel Prize.) Instead Einstein shifted strategies. He would no longer attack the theory as being incorrect. His new strategy was to absorb the quantum theory into his unified field theory. When the army of critics in Bohr's camp accused him of ignoring the quantum world, he fired back that his real goal was nothing short of cosmic in scope: to swallow up the quantum theory in its entirety in his new theory. Einstein used an analogy drawn from his own work. Relativity did not prove that Newtonian theory was completely wrong; it only showed that it was incomplete, that it could be subsumed into a larger theory. Thus, Newtonian mechanics is quite valid in its own particular domain: the realm of small velocities and large objects. Similarly, Einstein believed that the quantum theory's bizarre assumptions about cats being dead and alive simultaneously could be explained in a higher theory. In this respect, legions of Einstein's biographers have missed the point. Einstein's goal was not to prove the quantum theory incorrect, as many of his critics have claimed. He has too often been painted as the last dinosaur of classical physics, the aging rebel who found himself becoming the voice of reaction. Einstein's true goal was to expose the quantum theory's incompleteness and to use the unified field theory to complete it. In fact, one of the criteria for the unified field theory was that it reproduce the uncertainty principle in some approximation.

Einstein's strategy was to use general relativity and his unified field theory to explain the origin of matter itself, *to construct matter out of geometry.* In 1935, Einstein and Nathan Rosen investigated a novel way in which quantum particles such as the electron would emerge naturally as a consequence of his theory rather than as fundamental objects. In this way, he hoped to derive the quantum theory without ever having to face the problem of probabilities and chance. In most theories, elementary particles emerge as singularities, that is, regions where the equations blow up. Think of Newton's equations, for example, where the force is given by the inverse square of the distance between two objects. When this distance goes to zero, the force of gravity goes to infinity, giving us a singularity. Because Einstein wanted to derive the quantum theory from a deeper theory, he reasoned that he needed a theory totally free of singularities. (Examples of this exist in simple quantum theories. They are called "solitons" and resemble kinks in space; that is, they are smooth, not singular, and they can bounce off each other and maintain their same shape.)

Einstein and Rosen proposed a novel way to achieve such a solution. They started with two Schwarzschild black holes, defined on two parallel sheets of paper. By using scissors, one could cut out each black hole singularity and glue the two sheets back together. Thus, one obtains a smooth, singularity-free solution, which Einstein thought might represent a subatomic particle. Thus, *quantum particles can be viewed as tiny black holes.* (This idea was actually revived in string theory sixty years later, where there are mathematical relations that can turn subatomic particles into black holes and vice versa.)

This "Einstein-Rosen bridge," however, can be viewed in another way. It represents the first mention in the scientific literature of a "wormhole" that connects two universes. Wormholes are shortcuts through space and time, like a gateway or portal that connects two parallel sheets of paper. The concept of wormholes was introduced to the public by Charles Dodgson (otherwise known as Lewis Carroll), the Oxford mathematician and, most famously, the author of *Alice in Wonderland* and *Through the Looking Glass*. When Alice puts her hand through the Looking Glass, she is in effect entering a kind of Einstein-Rosen bridge connecting two universes—the strange world of Wonderland and the countryside of Oxford. It was realized, of course, that anyone who fell through an Einstein-Rosen bridge would be crushed to death by the intense gravitational force, enough to rip their atoms apart. Passage through the wormhole to a parallel universe was impossible if the black hole was stationary. (It would take another sixty years before the concept of wormholes would occupy a key role in physics.)

Eventually, Einstein gave up this idea, in part because he could not explain the richness of the subatomic world. He could not entirely explain all the curious properties of "wood" in terms of "marble." There were simply too many features of subatomic particles (e.g., mass, spin, charge, quantum numbers, etc.) that failed to emerge from his equations. His goal was to find the picture that would reveal the unified field theory in all its splendor, but one crucial problem was that not enough was known at that time about the properties of the nuclear force. Einstein was working decades before data from powerful atom smashers would clarify the nature of subatomic matter. As a result, the picture never came.

CHAPTER 8

War, Peace, and $E = mc^2$

In the 1930s, with the world caught in the vicelike grip of the Great Depression, chaos was once again stalking the streets of Germany. With the collapse of the currency, hard-working, middle-class citizens suddenly found their life savings wiped out almost overnight. The rising Nazi Party fed upon the misery and grievances of the German people, focusing their anger at the most convenient scapegoat, the Jews. Soon, with the backing of powerful industrialists, they became the strongest force in the Reichstag. Einstein, who had resisted the anti-Semites for years, realized that this time the situation was life-threatening. Although a pacifist, he was also realistic, adjusting his views in light of the meteoric rise of the Nazi Party. "This means that I am opposed to the use of force under any circumstances except when confronted by an enemy who pursues the destruction of life as an end in itself," he wrote. This flexibility would be put to the test.

In 1931, a book called *One Hundred Authorities against Einstein* was published, containing all kinds of anti-Semitic slander directed against the famous physicist. "The purpose of this publication is to oppose the terror of the Einsteinians with an account of the strength of their opposition," the document fumed. Einstein later quipped that they did not really need one hundred authorities to destroy relativity. If it were incorrect, one small fact would have been sufficient. In December 1932,

Einstein, unable to resist the rising tide of Nazism, left Germany for good. He told Elsa to look at their country house in Caputh and said sadly, "Turn around, you will never see it again." The situation deteriorated dramatically on January 30, 1933, when the Nazis, already the largest block in the Parliament, finally seized power, and Adolf Hitler was appointed as chancellor of Germany. The Nazis confiscated Einstein's property and his bank account, leaving him officially penniless, and took over his cherished Caputh vacation house, claiming to have found a dangerous weapon there. (It was later found to be a bread knife. The Caputh house was used during the Third Reich by the Nazi Bund Deutsches Mädel, the "League of German Girls"). On May 10, the Nazis held a public burning of banned books, Einstein's works among them. That year, Einstein wrote to the Belgian people, who were under the shadow of Germany: "Under today's conditions, if I were a Belgian, I would not refuse military service." His remarks were carried by the international media and earned him immediate scorn from both Nazis and fellow pacifists, many of whom believed that the only way to confront Hitler was with peaceful means. Einstein, realizing the true depths of the brutality of the Nazi regime, was unmoved: "The antimilitarists are falling on me as on a wicked renegade. . . . those fellows simply wear blinders."

Forced to flee Germany, Einstein the world traveler was once again a person without a home. On his trip to England in 1933, he stopped by to see Winston Churchill at his estate. Under "address" in Churchill's guest book, Einstein wrote, "None." Now near the top of the Nazi's hate list, he had to be careful of his personal security. A German magazine listing the enemies of the Nazi regime showed Einstein's picture on the front cover with the caption, "Not yet hanged." Anti-Semites were proud to say that if they could drive Einstein out of Germany, they could drive all Jewish scientists out. Meanwhile, the Nazis passed a new law requiring the dismissal of all Jewish officials, which was an immediate disaster for German physics. Nine Nobel laureates had to leave Germany because of the new civil service law, and seventeen hundred faculty members were dismissed in the first year, causing a vast hemorrhaging of German science and technology. The mass exodus out of Nazi-controlled Europe was staggering, virtually depleting the cream of the scientific elite.

Max Planck, ever the conciliator, refused all efforts by his colleagues to oppose Hitler publicly. He preferred to use private channels and even met personally with Hitler in May 1933, making one last final plea to prevent the collapse of German science. Planck would write, "I had hoped to convince him that he was doing enormous damage . . . by expelling our Jewish colleagues; to show how senseless and utterly immoral it was to victimize men who had always thought of themselves as Germans, and who had offered up their lives for Germany like everyone else." At that meeting, Hitler said that he had nothing against Jews, but they were all Communists. When Planck tried to reply, Hitler shouted back to him, "People say that I get attacks of nervous weakness, but I have nerves of steel!" He then slapped his knee and continued his tirade against Jews. Planck would regret, "I failed to make myself understood. . . . There is simply no language in which one can talk to such men."

Einstein's Jewish colleagues all fled Germany for their lives. Leo Szilard left with his life savings stuffed in his shoes. Fritz Haber fled Germany in 1933 for Palestine. (Ironically, as a loyal German scientist he had helped to develop poison gas for the German army, producing the notorious Zyklon B gas. Later, his own gas was used to kill many members of his family at the Auschwitz concentration camp.) Erwin Schrödinger, who was not Jewish, was also swept up by the hysteria. On March 31, 1933, when the Nazis declared a national boycott of all Jewish stores, he happened to be in front of Berlin's large Jewish department store, Wertheim's, when he suddenly witnessed gangs of storm troopers with Nazi swastikas beating up Jewish shopkeepers as the police and the crowd stood by and laughed. Schrödinger was incensed and went up to one of the storm troopers and berated him. Then the storm troopers turned and began to beat him instead. He could have been seriously hurt by this ferocious beating, but a young physicist wearing a Nazi swastika instantly recognized Schrödinger and was able to get him to safety. Badly shaken, Schrödinger would leave Germany for England and Ireland.

In 1943, the Nazis occupied Denmark, and Bohr, who was part Jewish, was targeted for extinction. He managed to escape just one step ahead of the Gestapo via neutral Sweden and then fly to Britain, although he almost died of suffocation on the plane because of an ill-fitting oxygen mask. Planck, a loyal patriot who never left Germany, also suffered horribly. His son was arrested for trying to assassinate Hitler, for which he was tortured by the Nazis and later executed.

Einstein, although in exile, was besieged with job offers from around the world. Leading universities in England, Spain, and France wished to capture this world-famous figure. Previously, he had been a guest professor at Princeton University. He had spent his winters in Princeton and his summers in Berlin. Abraham Flexner, representing a new institute to be formed at Princeton, largely with a five-million-dollar fund from the Bamberger fortune, had met several times with Einstein and approached him about the possibility of moving to the new institute. What appealed to Einstein was the fact that he would be free to travel and free of teaching duties. Although he was a popular lecturer, regularly breaking up audiences with his antics and enchanting royalty with amusing anecdotes, teaching and lecturing duties were taking time away from his beloved physics.

One colleague warned Einstein that coming permanently to the United States was like "committing suicide." The United States, before the sudden influx of Jewish scientists fleeing Nazi Germany, was considered a quiet backwater of science, with almost no institutions of higher learning capable of competing with Europe's. Defending his choice, Einstein wrote to Queen Elizabeth of Belgium, "Princeton is a wonderful little spot . . . a quaint ceremonious village of puny demigods on stilts. By ignoring certain special conventions I have been able to create for myself an atmosphere conducive to study and free of distraction." The news that Einstein had settled in the United States was heard around the world. The "pope of physics" had left Europe. The new Vatican would be the Institute for Advanced Study at Princeton.

When Einstein was shown his office for the first time, he was asked what he needed. Besides a desk and a chair, he said he needed a "large wastebasket . . . so I can throw away all my mistakes." (The institute also apparently made an offer to Erwin Schrödinger. But the latter, it is said, who was often accompanied by his wife and mistress and practiced an "open marriage" with a long list of lovers, found the atmosphere too stifling and conservative.) The American people were fascinated by the new arrival in New Jersey, who instantly became the country's most famous scientist. Soon, he was a familiar figure to all. Two Europeans, on a bet, sent a letter to "Dr. Einstein, America," to see if it would reach him. It did.

The 1930s were hard on Einstein personally. It seemed as if his worst fears about his son Eduard (fondly nicknamed Tedel) were confirmed when Eduard finally suffered a nervous breakdown in 1930 after a failed romance with an older woman. He was taken to the Burghozli psychiatric hospital in Zurich, the same hospital where Mileva's sister had been institutionalized. Diagnosed as schizophrenic, he was never to leave the care of an institution for the rest of his life except for short visits. Einstein, who always suspected that one of his sons might inherit mental problems from his wife, blamed "grave heredity." "I have seen it coming, slowly but irresistibly, ever since Tedel's youth," he wrote sadly. In 1933, his close friend Paul Ehrenfest, who helped to stimulate the early development of general relativity but suffered from depression, eventually committed suicide, shooting and killing his young mentally retarded son in the process.

After a prolonged, painful illness, Elsa, who had been with Einstein for about twenty years, died in 1936. According to friends, Einstein was "utterly ashen and shaken." Her death "severed the strongest tie he had with a human being." He took it hard but managed to slowly recuperate. He would write, "I have got used extremely well to life here. I live like a bear in my den. . . . The bearishness has been further enhanced by the death of my woman comrade, who was better with other people than I am."

After Elsa's death, he would live with his sister Maja, who had fled the Nazis; his stepdaughter Margot; and his secretary, Helen Dukas. He had started the final phase of his life. During the 1930s and 1940s he aged rapidly, and without Elsa to constantly harp about his appearance, the dashing, charismatic figure who dazzled kings and queens in his tuxedo reverted back into the old, bohemian ways of his youth. He now became the white-haired figure remembered most dearly by the public, the sage of Princeton, who would good-humoredly greet both children and royalty alike.

For Einstein, however, there was no rest. While at Princeton, he faced yet another challenge, the quest to build an atomic bomb. Back in 1905, Einstein had speculated that his theory might be able to explain how a small amount of radium could glow ferociously in the dark, its atoms releasing large quantities of power without apparent limit. In fact, the amount of energy locked in the nucleus could easily be a hundred million times greater than that stored in a chemical weapon. By 1920, Einstein had grasped the enormous practical implications of the energy locked in the nucleus of the atom when he wrote, "It might be possible, and it is not even improbable, that novel sources of energy of enormous effectiveness will be opened up, but this idea has no direct support from the facts known to us so far. It is very difficult to make prophecies, but it is within the realm of the possible." In 1921, he even speculated that at some point far in the future, the current economy, based on coal, might eventually be replaced by nuclear energy. But he also clearly understood two enormous problems. First, this cosmic fire could be used to forge an atomic bomb, with horrible consequences for humanity. He wrote prophetically, "All bombardments since the invention of firearms put together would be harmless child's play compared to its destructive effects." He also wrote that an atomic bomb could be used to unleash nuclear terrorism and even a nuclear war: "Assuming that it were possible to effect that immense energy release, we should merely find ourselves in an age compared to which our coal-black present would seem golden."

Last, and most important, he realized the enormous challenge in producing such a weapon. In fact, he doubted that it was doable in his lifetime. The practical problems of taking the terrible power locked in a single atom and magnifying it trillions of times was beyond anything possible in the 1920s. He wrote that it was as difficult "as firing at birds in the dark, in a neighborhood that has few birds."

Einstein realized that the key might be to somehow multiply the power of a single atom. If one could take the energy of an atom and then trigger the subsequent release of energy from nearby atoms, then one might be able to magnify this nuclear energy. He hinted that a chain reaction might happen if "the rays released . . . are in turn able to produce the same effects." But in the 1920s, he had no idea how such a chain reaction might be produced. Others, of course, also toyed with the idea of nuclear energy, not to benefit humanity, but for malevolent reasons. In April 1924, Paul Harteck and Wilhelm Groth informed the German Army Ordinance Department that "the country that exploits it first will have an incalculable advantage over the others."

The problem of releasing this energy is as follows: The nucleus of the atom is positively charged and hence repels other positive charges. Thus, the nucleus is protected against any random collisions that might unlock its nearly limitless energy. Ernest Rutherford, whose pioneering work led to the discovery of the nucleus of the atom, dismissed the atomic bomb, stating that "anyone who expects a source of power from the transformation of these atoms is talking moonshine." This stalemate was broken dramatically in 1932 when James Chadwick discovered a new particle, the neutron, a partner of the proton in the nucleus that is neutral in charge. If one could fire a beam of neutrons at the nucleus, then the neutron, undeterred by the electric field around the nucleus, might be able to shatter it, releasing nuclear energy. The thought occurred to physicists: a beam of these neutrons might effortlessly split the atom and trigger an atomic bomb.

While Einstein had doubts about the possibility of an atomic bomb, key events leading to nuclear fission were accelerating. In 1938, Otto Hahn and Fritz Strassmann of the Kaiser Wilhelm Institute for Physics in Berlin electrified the world of physics by splitting the uranium nucleus. They found traces of barium after bombarding uranium with neutrons, which indicated that the uranium nucleus split in half, creating barium in the process. Lise Meitner, a Jewish scientist and colleague of Hahn who had fled the Nazis, and her nephew Otto Frisch provided the missing theoretical basis to Hahn's result. Their results showed that the debris left over from the process weighed a bit less than the original uranium nucleus. It seemed as if mass was disappearing in this reaction. The splitting of the uranium atom also released 200 million electron volts of energy, which apparently appeared out of nowhere. Where did the missing mass go, and where did this energy mysteriously come from? Meitner realized that Einstein's equation $E = mc^2$ held the key to this puzzle. If one took the missing mass and multiplied it by c^2, then one found 200 million electron volts, precisely according to Einstein's theory. Bohr, when told of this startling verification of Einstein's equation, immediately grasped the significance of this result. He slapped his forehead and exclaimed, "Oh, what fools we all have been!"

In March of 1939, Einstein told the *New York Times* that the results so far "do not justify the assumption of a practical utilization of the atomic energies released in the process. . . . However, there is no single physicist with soul so poor who would allow this to affect his interest in this highly important subject." Ironically, that very same month, Enrico Fermi and Frédéric Joliot-Curie (Marie Curie's son-in-law) discovered that two neutrons can be released by the splitting of the uranium nucleus. This was a staggering result. If these two neutrons can go on to split two other uranium nuclei, then this would result in four neutrons, then eight, then sixteen, then thirty-two, ad infinitum, until the unimaginable power of the nuclear force was released in a chain reaction. Within a fraction of a second, the splitting of a single uranium atom could trigger the splitting of trillions upon trillions of other uranium atoms, releasing unimaginable quantities of nuclear energy. Fermi, looking out his window in Columbia University, mused grimly that a single atomic bomb could destroy all that he could see of New York City.

The race was on. Alarmed at the rapid speed of events, Szilard was worried that the Germans, who were leaders in atomic physics, would be the first to build an atomic bomb. In 1939, Szilard and Eugene Wigner drove to Long Island to visit Einstein to sign a letter that would be given to President Roosevelt.

The fateful letter, one of the most important in world history, began, "Some recent work by E. Fermi and L. Szilard, which has been communicated to me in manuscript, leads me to expect that the element uranium may be turned into a new and important source of energy in the immediate future." Ominously, the letter noted that Hitler had invaded Czechoslovakia and had sealed off the Bohemian pitchblende mines, a rich source of uranium ore. And then the letter warned, "A single bomb of this type, carried by boat or exploded in a port, might well destroy the whole port with some of the surrounding territory. However, such bombs might very well prove to be too heavy for transportation by air." Alexander Sachs, a Roosevelt advisor, was given the letter to pass onto the president. When Sachs asked Roosevelt if he understood the extreme gravity of the letter, Roosevelt replied, "Alex. What you are after is to see that the Nazis don't blow us up." He turned to General E. M. Watson and said, "This requires action." Only six thousand dollars was approved for the whole year's research on uranium. However, interest in the atomic bomb was given a sudden boost when the secret Frisch-Peierls report reached Washington in the fall of 1941. British scientists, working independently, confirmed all the details outlined by Einstein, and on December 6, 1941, the Manhattan Engineering Project was secretly set up.

Under the direction of J. Robert Oppenheimer, who had worked on Einstein's theory of black holes, hundreds of the world's top scientists were secretly contacted and then shipped out to Los Alamos in the desert of New Mexico. At every major university, scientists like Hans Bethe, Enrico Fermi, Edward Teller, and Eugene Wigner quietly left after receiving a tap on the shoulder. (Not everyone was pleased by the intense interest in the atomic bomb. Lise Meitner, whose work helped to trigger the project, staunchly refused to be part of any work on the bomb. She was the only prominent Allied nuclear scientist to refuse the call to join the group at Los Alamos. "I will have nothing to do with a bomb!" she stated flatly. Years later, when Hollywood scriptwriters tried to glamorize her in the film *The Beginning of the End*, as the woman who bravely smuggled out the blueprint for the bomb as she fled Nazi Germany, she replied, "I would rather walk naked down Broadway" than be part of this fanciful, scurrilous effort.)

Einstein was aware that all his close colleagues at Princeton were suddenly disappearing, leaving a mysterious mailing address in Santa Fe, New Mexico. Einstein himself, though, was never given the tap on the shoulder and sat out the entire war at Princeton. The reason for this has been revealed in declassified war documents. Vannevar Bush, the chief of the Office of Scientific Research and Development and Roosevelt's trusted advisor, wrote, "I wish very much that I could place the whole thing before him [Einstein] . . . but this is utterly impossible in view of the attitude of people here in Washington who have studied his whole history." FBI and army intelligence concluded that Einstein could not be trusted: "In view of his radical background, this office would not recommend the employment of Dr. Einstein, on matters of a secret nature, without a very careful investigation, as it seems unlikely that a man of his background could, in such a short time, become a loyal American citizen." Apparently, the FBI did not realize that Einstein was already well aware of the project and in fact had helped to set it into motion in the first place.

Einstein's FBI file, recently declassified, runs 1,427 pages. J. Edgar Hoover had targeted Einstein as being either a Communist spy or a dupe at best. The agency carefully screened every piece of gossip about him and filed it away. Ironically, the FBI was curiously negligent in confronting Einstein himself, as if they feared him. Instead, agents preferred to interview and harass those surrounding him. As a result, the FBI became a repository of hundreds of letters from every crank and paranoid. In particular, they filed away reports that Einstein was working on some kind of death ray. In May 1943, a navy lieutenant called on Einstein, asking him if he would be willing to work on weapons and high explosives for the U.S. Navy. "He felt very bad about being neglected. He had not been approached by anyone to do any war work," wrote the lieutenant. Einstein, always quick with a quip, remarked that he was now in the navy without having to get a haircut.

The intense Allied effort to build an atomic bomb was stimulated by fears of the German bomb. In reality, the German war effort was badly understaffed and underfunded. Werner Heisenberg, Germany's greatest quantum physicist, was put in charge of a team of scientists to work on the German project. In the fall of 1942, when German scientists realized that it would take another three years of strenuous effort to produce an atomic bomb, Albert Speer, the Nazi armaments minister, decided to temporarily shelve the project. Speer made a strategic error, assuming that Germany would win the war in three years, making the bomb unnecessary. Nevertheless, he continued funding research on nuclear-powered submarines.

Heisenberg was hampered by other problems. Hitler declared that ordinance development would proceed only on weapons that promised results in six months, an impossible deadline. In addition to a lack of funding, German laboratories were under attack by Allied forces. In 1942, a commando squad successfully blew up Heisenberg's heavy-water factory in Vemork, Norway. In contrast to Fermi's decision to build a carbon-based reactor, the Germans chose to build a heavy-water reactor that could use natural uranium, which was plentiful, rather than the extremely rare uranium-235. In 1943, the Allies hit Berlin hard with saturation bombing, forcing Heisenberg to move his laboratory. The Kaiser Wilhelm Institute for Physics was evacuated to Hechingen, in the hills south of Stuttgart. Heisenberg had to build Germany's reactor in a rock cellar in nearby Haigerloch. Under intense pressure and bombing, they never succeeded in sustaining a chain reaction.

Meanwhile, physicists in the Manhattan Project were rushing to process enough plutonium and uranium for four atomic bombs. They were doing calculations right up to the time of the fateful detonation in Alamogordo, New Mexico. The first bomb, based on plutonium-239, was detonated in July 1945. After the decisive Allied victory over the Nazis, many physicists thought that the bomb would be unnecessary against the remaining enemy, Japan. Some believed that a demonstration atomic bomb should be detonated on a deserted island, witnessed by a delegation of Japanese officials, to warn the Japanese that surrender was inevitable. Others even drafted a letter to President Harry Truman asking him not to drop the bomb on Japan. Unfortunately, this letter was never delivered. One scientist, Joseph Rotblatt, even resigned from the atomic bomb project, stating that his work was finished and that the bomb should never be used against Japan. (He would later win the Nobel Prize for peace.)

Nevertheless, the decision was made to drop not one, but two atomic bombs on Japan in August 1945. Einstein was vacationing at Saranac Lake in New York. That week Helen Dukas heard the news on the radio. She recalled that the report "said a new kind of bomb has been dropped on Japan. And then I knew what it was because I knew about the Szilard thing in a vague way. . . . As Professor Einstein came down to tea, I told him, and he said, 'Oh, Weh' [Oh my God]."

In 1946, Einstein made the cover of *Time*. Ominously, this time there was a nuclear fireball erupting behind him. The world suddenly realized that the next war, World War III, might be fought with atomic bombs. But, Einstein noted, because nuclear weapons might send civilization back thousands of years, World War IV would be fought with rocks. That year, Einstein became chairman of the Emergency Committee of Atomic Scientists, perhaps the first major antinuclear organization, and used it as a platform to argue against the continued building of nuclear weapons—and to advocate one of his cherished causes, world government.

Meanwhile, amidst the storm unleashed by the atomic and hydrogen bombs, Einstein maintained his peace and sanity by returning stubbornly to his physics. In the 1940s, pioneering work was still being done in the areas he helped to found, including cosmology and the unified field theory. This was to be his last and final attempt to "read the mind of God."

After the war, Schrödinger and Einstein maintained a lively transatlantic correspondence. Almost alone, these two fathers of the quantum theory resisted the tide of quantum mechanics and focused on the quest for unification. In 1946, Schrödinger confessed to Einstein, "You are after big game. You are on a lion hunt, while I am speaking of rabbits." Encouraged by Einstein, Schrödinger continued his hot pursuit of a particular type of unified field theory, called "affine field theory." Soon, Schrödinger completed his own theory, which convinced him that he finally accomplished what Einstein had failed to achieve, the unification of light and gravity. He marveled that his new theory was a "miracle," a "totally unhoped for gift from God."

Working in Ireland, Schrödinger had felt isolated from the mainstream of physics, being reduced to a college administrator and a has-been. Now he was convinced that his new theory might win him a second Nobel Prize. Hurriedly, he called a major press conference. Ireland's prime minister, Eamon De Valera, and others showed up to listen to his presentation. When a reporter asked him how confident he was about his theory, he said, "I believe I am right. I shall look an awful fool if I am wrong." However, Einstein quickly saw that Schrödinger had pursued a theory that he himself had discarded years earlier. As physicist Freeman Dyson wrote, the trail leading to the unified field theory is littered with the corpses of failed attempts.

Undaunted, Einstein kept working on the unified field theory, largely in isolation from the rest of the physics community. Lacking a guiding physical principle, he would try to find beauty and elegance in his equations. As mathematician G. H. Hardy once said, "Mathematical patterns like those of the painters or the poets must be beautiful. The ideas, like the colors or the words must fit together in a harmonious way. Beauty is the first test. There is no permanent place for ugly mathematics." But lacking something like an equivalence principle for the unified field theory, Einstein was left without a guiding star. He lamented the fact that other physicists did not see the world as he did, but he never lost any sleep over this. He would write, "I have become a lonely old fellow. A kind of patriarchal figure who is known chiefly because he does not wear socks and is displayed on various occasions as an oddity. But in my work I am more fanatical than ever and I really entertain the hope that I have solved my old problems of the unity of the physical field. It is, however, like being in an airship in which one can cruise around in the clouds but cannot see clearly how one can return to reality, i.e. to earth."

Einstein realized that by working on his unified field theory rather than the quantum theory, he was isolated from the main avenues of research at the institute. "I must seem like an ostrich who forever buries its head in the relativistic sand in order not to face the evil quanta," he lamented. Over the years, other physicists would whisper that he was over the hill and behind the times, but this did not bother him. "I am generally regarded as a sort of petrified object, rendered blind and deaf by the years. I find this role not too distasteful, as it corresponds very well with my temperament," he wrote.

In 1949, on his seventieth birthday, a special celebration was held in Einstein's honor at the institute. Scores of physicists came to praise the greatest scientist of their time and contribute articles for a book in his honor. However, from the tone of some speakers and interviews with the press, it became apparent that some of them took Einstein to task for his position on the quantum theory. Einstein partisans were not happy with this, but Einstein took it good-naturedly. A family friend, Thomas Bucky, noted that "Oppenheimer made fun of Einstein in a magazine article with such statements as, 'He's old. Nobody pays any attention to him anymore.' We were madder than all hell about it. But Einstein was not mad at all. He just didn't believe it and later Oppenheimer denied he had said it."

That was Einstein's manner, to take his critics with a grain of salt. When the book in his honor came out, he wrote in good humor, "This is not a jubilee book for me, but an impeachment." He was a seasoned enough scientist to know that new ideas were hard to come by, and that he was not producing ideas like he did in his youth. As he would write, "Anything really new is invented only in one's youth. Later one becomes more experienced, more famous—and more stupid."

What kept him going, however, were the clues he saw everywhere that unification was one of the grand schemes of the universe. He would write, "Nature shows us only the tail of the lion. But I do not doubt that the lion belongs to it even though he cannot at once reveal himself because of his enormous size." Every day, when he woke up, he would ask himself a simple question: If he were God, how would he create the universe? In fact, given all the constraints necessary to create a universe, he asked himself another question: Did God have any choice? As he gazed at the universe, everything he saw told him that unification was the greatest theme in nature, that God could not have created a universe that made gravity, electricity, and magnetism as separate entities. What he lacked, as he knew, was a guiding principle, a physical picture that would light the way to the unified field theory. None came.

With special relativity the picture was a sixteen-year-old youth racing after a light beam. With general relativity, it was a man leaning back in his chair, about to fall, or marbles rolling on curved space. However, with the unified field theory, he had no such guidance. Einstein was famous for his statement, "Subtle is the Lord, but malicious he is not." After he struggled for so many decades on the problem of unification, he admitted to his assistant Valentine Bargman, "I have second thoughts. Maybe God *is* malicious."

Although the quest for a unified field theory was known to be the hardest problem in all of physics, it was also the most glamorous and seduced legions of physicists. It is ironic, for example, that Wolfgang Pauli, one of Einstein's severest critics of the unified field theory, would eventually catch the bug himself. In the late 1950s, both Heisenberg and Pauli were increasingly interested in a version of the unified field theory that they claimed could solve the problems that stumped Einstein for thirty years. In fact, writes Pais, "From 1954 to the end of his life, Heisenberg (d. 1976) was immersed in attempts at deriving all of particle physics from a fundamental non-linear wave equation." In 1958, Pauli visited Columbia University and gave a presentation on the Heisenberg-Pauli version of the unified field theory. The audience, needless to say, was skeptical. Niels Bohr, who was in the audience, finally stood up and said, "We in the back are convinced your theory is crazy. But what divides us is whether your theory is crazy enough."

Physicist Jeremy Bernstein, also in the audience, remarked, "It was an uncanny encounter of two giants of modern physics. I kept wondering what in the world a non-physicist visitor would have made of it." Eventually, Pauli became disillusioned with the theory, believing it had too many flaws. When his collaborator insisted on plunging ahead with the theory, Pauli wrote to Heisenberg and enclosed a blank sheet of paper, stating that if his theory was really the unified field theory, then this blank sheet of paper was a work by Titian.

Although progress in the unified field theory was slow and painful, there were plenty of other interesting breakthroughs that kept Einstein busy. One of the strangest was time machines.

To Newton, time was like an arrow. Once fired, it unerringly flew in a straight line, never deviating from its path. One second on the earth was one second in outer space. Time was absolute and beat uniformly through the entire universe at the same rate. Events could take place simultaneously throughout the universe. However, Einstein introduced the concept of relative time, so one second on the earth was not one second on the moon. Time was like Old Man River, meandering its way past planets and stars, slowing down as it went by neighboring heavenly bodies. The question that the mathematician Kurt Gödel now raised was, can the river of time have whirlpools and turn back on itself? Or can it fork into two rivers, creating a parallel universe? Einstein was forced to confront this question in 1949 when Gödel, Einstein's neighbor at the institute and arguably the greatest mathematical logician of the century, showed that Einstein's equations allowed for time travel. Gödel started with a universe that was filled with a gas and rotating. If one started off in a rocket ship and went around the entire universe, then one might arrive on the earth before one left! In other words, time travel would be a natural phenomenon in Gödel's universe, where one would routinely travel back in time during a trip around the universe.

This shook Einstein. So far, every time people tried to find solutions to Einstein's equations, they found solutions that seemed to fit the data. The perihelion of Mercury, the red shift, the bending of starlight, the gravity of a star, and so on, all fit experimental data very nicely. Now, his equations were giving solutions that challenged all our beliefs about time. If time travel were routinely possible, then history could never be written. The past, like shifting sands, could be changed anytime someone entered his or her time machine. Worse, one might destroy the universe itself by creating a time paradox. What if you went back in time and shot your parents before you were born? This was problematic, because how could you be born in the first place if you just killed your parents?

Time machines violated causality, which was a cherished principle of physics. Einstein was not happy with the quantum theory precisely because it replaced causality with probabilities. Now, Gödel was eliminating causality entirely! After much consideration, Einstein finally dismissed Gödel's solution by pointing out that it did not fit the observational data: the universe was expanding, not rotating, so time travel, at least for the time being, could be dismissed. But this left open the possibility that if the universe rotated instead of expanded, then time travel would be routine. It would, however, take another five decades before the concept of time travel would be revived into a major field of investigation.

The 1940s was also a turbulent time in cosmology. George Gamow, who was Einstein's liaison with the U.S. Navy during the war, was less interested in designing explosives than asking questions about the biggest explosion of all, the big bang. Gamow would ask himself several questions that would turn cosmology upside down. He took the big bang theory to its logical conclusion. He shrewdly speculated that if the universe was indeed born in a fiery explosion, then it should be possible to detect the leftover heat from the early fireball. There should be an "echo of creation" from the big bang itself. He used the work of Boltzmann and Planck, who showed that the color of a hot object should correlate with its temperature since both are different forms of energy. For example, if an object is red hot, it means that its temperature is approximately 3,000 degrees Celsius. If an object is yellow hot (like our sun), then it is roughly 6,000 degrees Celsius (which is the temperature of the surface of our sun). Similarly, our own bodies are warm, so we can calculate the "color" of our bodies, which correlates to infrared radiation. (Army night-vision goggles are effective because they detect the infrared radiation emitted from our warm bodies.) Arguing that the Big Bang happened billions of years ago, two members of Gamow's group, Robert Herman and Ralph Alpher, calculated as early as 1948 that the afterglow of the Big Bang should be 5 degrees above absolute zero, which is remarkably close to the correct value. This radiation corresponds to microwave radiation. Therefore, the "color of creation" is microwave radiation. (This microwave radiation, which was eventually found decades later and determined to correspond to 2.7 degrees above absolute zero, would completely revolutionize the field of cosmology.)

Although he was relatively isolated at Princeton, Einstein lived to see the day when his theory of general relativity was spawning rich new avenues of research in cosmology, black holes, gravity waves, and other areas. However, the last years of his life were also filled with sorrows. In 1948, he received word that Mileva, after a long, hard life caring for their mentally ill son, had passed away, apparently of a stroke during a tantrum of Eduard's. (Later, 85,000 francs in cash was found stuffed in her bed, apparently the last money left from her Zurich apartments. It was used to help pay for Eduard's long-term care.) In 1951, his dear sister Maja died.

In 1952, Chaim Weizmann, the man who had organized Einstein's triumphant tour of America in 1921, passed away after being president of Israel. Unexpectedly, Israel's premier, David Ben-Gurion, then offered Einstein the presidency of Israel. Although it was quite an honor, he had to decline.

In 1955, Einstein received word that Michele Besso, who had helped Einstein refine his ideas on special relativity, had died. In a letter to Besso's son, Einstein wrote movingly, "What I admired most about Michele was the fact that he was able to live so many years with one woman, not only in peace but also in constant unity, something I have lamentably failed at twice. . . . So in quitting this strange world he has once again preceded me by a little. That doesn't mean anything. For those of us who believe in physics, this separation between past, present, and future is only an illusion, however tenacious."

That year, with his health failing, he said, "It is tasteless to prolong life artificially. I have done my share; it is time to go. I will do it elegantly." Einstein finally died on April 18, 1955, of a burst aneurysm. After his death, the cartoonist Herblock published in the *Washington Post* a moving cartoon depicting the earth, as seen from outer space, with a large sign that read, "Albert Einstein lived here." That night, newspapers around the world flashed over the wire services a photograph of Einstein's desk. On it was the manuscript for his greatest unfinished theory, the unified field theory.

CHAPTER 9

Einstein's Prophetic Legacy

Most biographers uniformly ignore the last thirty years of Einstein's life, considering it almost an embarrassment unworthy of a genius, a stain on his otherwise sterling history. However, scientific developments in the last few decades have given us an entirely new look into Einstein's legacy. Because his work was so fundamental, reshaping the very foundations of human knowledge, his impact continues to reverberate throughout physics. Many of the seeds planted by Einstein are now germinating in the twenty-first century, mainly because our instruments such as space telescopes, X-ray space observatories, and lasers are now powerful and sensitive enough to verify a variety of his predictions made decades ago.

In fact, crumbs that have tumbled off Einstein's plate are now winning Nobel Prizes for other scientists. Furthermore, with the rise of superstring theory, Einstein's concept of unification of all forces, once the subject of derision and derogatory comments, is now assuming center stage in the world of theoretical physics. This chapter discusses new developments in three areas where Einstein's enduring legacy continues to dominate the world of physics: the quantum theory, general relativity and cosmology, and the unified field theory.

When Einstein first wrote his paper on Bose-Einstein condensation in 1924, he did not believe that this curious phenomenon would be discovered anytime soon. One would have to cool materials down to near absolute zero before all the quantum states could collapse into a giant superatom.

In 1995, Eric A. Cornell from the National Institute of Standards and Technology and Carl E. Weiman of the University of Colorado did just that, producing a pure Bose-Einstein condensate of 2,000 rubidium atoms at twenty-billionths of a degree above absolute zero. In addition, Wolfgang Ketterle of MIT independently produced Bose-Einstein condensates with enough sodium atoms to do important experiments on them, such as proving that these atoms displayed interference patterns consistent with atoms that were coordinated with each other. In other words, they acted like the superatom predicted by Einstein over seventy years earlier.

Since the initial announcement, discoveries in this fast-moving field have come rapidly. In 1997, Ketterle and his colleagues at MIT created the world's first "atom laser" using Bose-Einstein condensates. What gives laser light its marvelous properties is the fact that the photons march in unison and lockstep with each other, while ordinary light is chaotic and incoherent. Since matter also has wavelike properties, physicists speculated that beams of atoms could also be made to "lase" as well, but the lack of Bose-Einstein condensates hindered progress in this direction. These physicists accomplished their feat by first cooling down a collection of atoms until they condensed. Then they hit the condensate with a laser beam, which turned the atoms into a synchronized beam.

In 2001, Cornell, Weiman, and Ketterle were awarded the Nobel Prize in physics. The Nobel Prize committee cited them "for the achievement of Bose-Einstein condensation in dilute gases of alkali atoms, and for early fundamental studies of the properties of the condensates." The practical applications of Bose-Einstein condensates are just being realized. These beams of atomic lasers could prove valuable in the future when applied to nanotechnology. They may allow the manipulation of individual atoms and the creation of layers of atomic films for semiconductors in computers of the future.

In addition to atomic lasers, some physicists have speculated that quantum computers (computers that compute on individual atoms) could be based on Bose-Einstein condensates, which could eventually replace silicon-based computers. Others have speculated that dark matter, in part, could be composed of Bose-Einstein condensates. If so, then this obscure state of matter could make up most of the universe.

Einstein's contributions have also forced quantum physicists to rethink their devotion to the original Copenhagen interpretation of the theory. Back in the 1930s and 1940s, when quantum physicists were snickering behind Einstein's back, it was easy to ignore this giant of physics because so many discoveries in quantum physics were being made almost daily. Who had time to contemplate the foundations of the quantum theory when physicists were scrambling to collect Nobel Prizes like apples picked off a tree? Hundreds of calculations on the properties of metals, semiconductors, liquids, crystals, and other materials could now be performed, each of which might create entire industries. There was simply no time to spare. As a consequence, physicists for decades simply got used to the Copenhagen school, brushing the unanswered deeper philosophical questions under the rug. The Bohr-Einstein debates were forgotten. However, now that many of the "easy" questions about matter have been picked clean, the much more difficult questions raised by Einstein are still unanswered. In particular, scores of international conferences are taking place around the world as physicists re-examine the cat problem mentioned in chapter 7. Now that experimentalists can manipulate individual atoms, the cat problem is no longer just an academic question. In fact, the ultimate fate of computer technology, which accounts for a large fraction of the world's wealth, may depend on its resolution since computers of the future may use transistors made of individual atoms.

Of all the alternatives, the Copenhagen school of Bohr is now recognized to have the least attractive answer to the cat problem, although there has been no experimental deviation from Bohr's original interpretation. The Copenhagen school postulates that there is a "wall" that separates the commonsense, macroscopic world of trees, mountains, and people that we see around us, from the mysterious, nonintuitive microscopic world of the quantum and waves. In the microscopic world, subatomic particles exist in a nether state between existence and nonexistence. However, we live on the other side of the wall, where all wave functions have collapsed, so our macroscopic universe seems definite and well defined. In other words, there is a wall separating the observer from the observed.

Some physicists, including Nobel laureate Eugene Wigner, even went further. The key element of observing, he stressed, is consciousness. It takes a conscious observer to make an observation and determine the reality of the cat. But who observes the observer? The observer must also have another observer (called "Wigner's friend") to determine that the observer is alive. But this implies an infinite chain of observers, each one observing the other, each one determining that the previous observer is alive and well. To Wigner, this meant that perhaps there was a cosmic consciousness that determined the nature of the universe itself! As he said, "The very study of the external world led to the conclusion that the content of the consciousness is the ultimate reality." Some have argued therefore that this proves the existence of God, some sort of cosmic consciousness, or that the universe itself was somehow conscious. As Planck once said, "Science cannot solve the ultimate mystery of Nature. And it is because in the last analysis we ourselves are part of the mystery we are trying to solve."

Over the decades, other interpretations have been proposed.

In 1957, Hugh Everett, then a graduate student of physicist John Wheeler, proposed perhaps the most radical solution to the cat problem, the "many worlds" theory, stating that all possible universes exist simultaneously. The cat could indeed be dead and alive simultaneously, because the universe itself has split into two universes. The implications of this idea are quite unsettling, because it means that the universe is constantly bifurcating at each quantum instant, spinning off into infinite numbers of quantum universes. Wheeler himself, originally enthusiastic about his student's approach, later abandoned it, stating that it carried too much "metaphysical baggage." For example, imagine a cosmic ray that penetrates Winston Churchill's mother's womb, triggering a miscarriage. Thus, one quantum event separates us from a universe in which Churchill never lived to rally the people of England and the world against the murderous forces of Adolf Hitler. In that parallel universe, perhaps the Nazis won World War II and enslaved much of the world. Or imagine a world where a solar wind, triggered by quantum events, pushed a comet or meteor from its path 65 million years ago, so it never hit the Yucatan Peninsula of Mexico and never wiped out the dinosaurs. In that parallel universe, humans never emerged and Manhattan, where I am now living, is populated by rampaging dinosaurs.

The mind is sent spinning contemplating all possible universes. After decades of futile arguing over various interpretations of the quantum theory, in 1965 John Bell, a physicist at the nuclear laboratory at CERN in Geneva, Switzerland, analyzed an experiment that would decisively prove or disprove Einstein's criticism of the quantum theory. This would be the acid test. He was sympathetic to the deep philosophical questions raised by Einstein decades earlier, and proposed a theorem that would finally settle the question. (Bell's theorem is based on re-examining a variation of the old EPR experiment and analyzing the correlation between the two particles moving in opposite directions.) The first credible experiment was performed in 1983 by Alain Aspect at the University of Paris, and the results confirmed the quantum mechanical viewpoint. Einstein was wrong about his criticism of the quantum theory.

But if Einstein's criticism of the quantum theory could now be ruled out, then which of the various quantum mechanical schools is correct? Most physicists today believe that the Copenhagen school is woefully incomplete. Bohr's wall separating the microscopic world from the macroscopic world does not seem valid in today's world, when we can now manipulate individual atoms. "Scanning tunneling microscopes" can in fact displace individual atoms and have been used to spell out "IBM" and create a working abacus made of atoms. In addition, a whole new field of technology, called "nanotechnology," has been created based on the manipulation of atoms. Experiments like Schrödinger's cat experiment can now be performed on individual atoms.

Despite this, there is still no solution to the cat problem that is satisfactory to all physicists. Almost eighty years since Bohr and Einstein clashed at the Solvay Conference, however, some leading physicists, including several Nobel laureates, have converged on the idea of "decoherence" to resolve the cat problem. Decoherence starts with the fact that the wave function of a cat is quite complicated because it contains something on the order of 10^{25} atoms, a truly astronomical number. Hence the interference between the live cat's wave and the dead cat's wave is quite intense. This means that the two wave functions can coexist simultaneously in the same space but can never influence each other. The two wave functions have "decohered" from each other and no longer sense each other's presence. In one version of decoherence, wave functions never "collapse," as claimed by Bohr. They simply separate and, for all intents and purposes, never interact again.

Nobel laureate Steven Weinberg compares this to listening to a radio. By turning a dial, we can tune successively to many radio stations. Each frequency has decohered from the others, so there is no interference between stations. Our room is simultaneously filled with signals from all radio stations, each one yielding an entire world of information, yet they do not interact with each other. And our radio tunes into only one at a time.

Decoherence sounds attractive, since it means that ordinary wave theory can be used to resolve the problem of the cat without resorting to the "collapse" of the wave function. In this picture, waves never collapse. However, the logical conclusions are disturbing. In the final analysis, decoherence implies a "many worlds" interpretation. But instead of radio stations that do not interfere, now we have entire universes that do not interact. It may seem strange, but this means that sitting in the very room where you are reading this book, there exist the wave function of parallel worlds where the Nazis won World War II, where people speak in strange tongues, where dinosaurs battle in your living room, where alien creatures walk the earth, or where the earth never existed in the first place. Our "radio" is tuned only to the familiar world we live in, but within this room there exist other "radio stations" where insane, bizarre worlds coexist with ours. We cannot interact with these dinosaurs, monsters, and aliens walking in our living rooms because we live on a different "radio" frequency and have decohered from them. As Nobel laureate Richard Feynman has said, "I think I can safely say that nobody understands quantum mechanics."

While Einstein's critique of the quantum theory helped to sharpen its development but may not have brought forth a wholly satisfactory solution to its paradoxes, his ideas have been vindicated elsewhere, most spectacularly in general relativity. In an era of atomic clocks, lasers, and supercomputers, scientists are mounting the kind of high-precision tests of general relativity that Einstein could only dream about. In 1959, for example, Robert V. Pound and G. A. Rebka of Harvard finally confirmed

Einstein's prediction of gravitational red shift in the laboratory, that is, that clocks beat at different rates in a gravitational field. They took radioactive cobalt and shot radiation from the basement of Lyman Laboratory at Harvard to the roof, 74 feet above. Using an extremely fine measuring device (which used the Mossbauer effect), they showed that photons lost energy (hence were reduced in frequency) as they made the journey to the top of the laboratory. In 1977, astronomer Jesse Greenstein and his colleagues analyzed the beating of time in a dozen white dwarf stars. As expected, they confirmed that time slowed down in a large gravitational field.

The solar eclipse experiment has also been redone with extreme precision on a number of occasions. In 1970, astronomers pinpointed the location of two extremely distant quasars, 3C 279 and 3C 273. The light from these quasars bent as predicted by Einstein's theory.

The introduction of atomic clocks also has revolutionized the way in which precision tests can be performed. In 1971, atomic clocks were placed on a jet plane, which was flown both East to West and West to East. These atomic clocks, in turn, were then compared with atomic clocks that were stationary at the Naval Observatory in Washington, D.C. By analyzing the atomic clocks on the jets traveling at different velocities (but with constant altitude), scientists could verify special relativity. Then, by analyzing jets traveling at the same speed but different altitude, they could test the prediction of general relativity. On both occasions, the results verified Einstein's predictions, within experimental error.

The launching of space satellites has also revolutionized the way in which general relativity can be tested. The Hipparcos satellite, launched by the European Space Agency in 1989, spent four years calculating the deflection of starlight by the sun, even analyzing stars that are 1,500 times fainter than the stars in the Big Dipper. In deep space, there is no necessity to wait for an eclipse, and experiments can be conducted all the time. Without fail, they found that starlight bent according to Einstein's prediction. In fact, they found that starlight from halfway across the sky was bent by the sun.

In the twenty-first century, a variety of other precision experiments are planned to test the precision of general relativity, including more experiments on double stars and even bouncing laser signals off the moon. But the most interesting precision tests may come from gravity waves. Einstein predicted gravity waves in 1916. However, he despaired of ever being able to see confirmation of these elusive phenomena in his lifetime. The experimental equipment of the early twentieth century was simply too primitive. But in 1993, the Nobel Prize was awarded to two physicists, Russell Hulse and Joseph Taylor, for indirectly verifying the existence of gravity waves by examining double stars rotating around each other.

They examined PSR 1913+16, a double neutron star about 16,000 light-years from Earth, in which two dead stars orbit each other every seven hours and forty-five minutes, releasing copious quantities of gravity waves in their wake. Imagine, for example, stirring a pot of molasses with two spoons, each spoon rotating around the other. As each spoon moves in the molasses, it leaves a trail of molasses in its wake. Similarly, if we replace the molasses with the fabric of space-time and the spoons by dead stars, we find two stars chasing each other in space, emitting waves of gravity. Since these waves carry energy, the two stars eventually lose energy and gradually spiral together. By analyzing the signals of this double-star system, one can experimentally calculate the precise decay in the orbit of the double star. As expected from Einstein's general relativity theory, the two stars come closer by a millimeter every revolution. Over a year, the separation of the stars decreases by a yard in an orbit that is 435,000 miles in diameter, which is precisely the number that can be calculated from Einstein's equations. In fact, the two stars will completely collapse in 240 million years owing to the loss of gravity waves. This precision experiment can be reinterpreted as a way in which to test the accuracy of Einstein's general relativity. The numbers are so precise that we can conclude that general relativity is 99.7% accurate (well within experimental error).

More recently, there is intense interest in a series of far-reaching experiments to observe gravity waves directly. The LIGO (Laser Interferometer Gravitational Wave Observatory) project may soon be the first to observe gravitational waves, perhaps from black holes colliding in outer space. LIGO is a physicist's dream come true, the first apparatus powerful enough to measure gravity waves. LIGO consists of three laser facilities in the United States (two in Hanford, Washington, and one in Livingston, Louisiana). It is actually one part of an international consortium, including the French-Italian detector called VIRGO in Pisa, Italy; a Japanese detector called TAMA outside Tokyo; and a British-German detector called GEO600 in Hanover, Germany. Altogether, LIGO's final construction cost will be $292 million (plus $80 million for commissioning and upgrades), making it the most expensive project ever funded by the National Science Foundation.

The laser detectors used in LIGO look very much like the device used by Michelson-Morley at the turn of the century to detect the aether wind, except that laser beams are used instead of ordinary light beams. A laser beam is split into two separate beams that move perpendicular to each other. After hitting a mirror, these two beams are then reunited. If a gravity wave were to hit the interferometer, there would be a disturbance in the lengths of the paths of the laser beams, which could be seen as an interference pattern between the two beams. To make sure that the signal hitting the laser apparatus is not a spurious one, laser detectors are required to be distributed around the planet Earth. Only a huge gravity wave much bigger than Earth would be able to fire the detectors all at once.

Eventually, a series of these laser detectors will be placed in outer space by NASA and the European Space Agency. Around 2010, NASA will launch three satellites, called LISA (Laser Interferometry Space Antenna). They will orbit around the sun at approximately the same distance as the earth's orbit. The three laser detectors will form an equilateral triangle in outer space (about 3 million miles on a side). The system will be so delicate it will be able to detect vibrations of one part in a billion trillion (corresponding to a shift that is one-hundredth the width of a single atom), allowing scientists to detect the original shock waves from the big bang itself. If all goes well, LISA should be able to peer to within the first trillionth of a second after the big bang, making it perhaps the most powerful of all cosmological tools to exploring creation. This is essential, because it is believed that LISA may be able to find the first experimental data on the precise nature of the unified field theory, the theory of everything.

Yet another important tool introduced by Einstein was gravity lenses. Back in 1936, he proved that nearby galaxies can act as gigantic lenses that focus the light from distant objects. It would take many decades for these Einstein lenses to be observed. The first breakthrough came in 1979, when astronomers observed the quasar Q0957+561 and found that space was being warped and acting as a lens to concentrate light.

In 1988, the first observation of an Einstein ring was from the radio source MG1131+0456, and about twenty, mostly fragments of rings, have been observed since then. In 1997, the first completely circular Einstein rings were observed with the Hubble Space Telescope and Britain's MERLIN (Multi-Element Radio Linked Interferometer Network) radio telescope array. By analyzing the distant galaxy 1938+666, they found the characteristic ring that surrounded the galaxy. "At first sight, it looked artificial and we thought it was some sort of defect in the image, but then we realized we were looking at a perfect Einstein ring!"

said Dr. Ian Brown of the University of Manchester. Astronomers in Britain were elated by the discovery, declaring, "It's a bulls-eye!" The ring is tiny. It is only a second of an arc, or roughly the size of a penny viewed from a distance of two miles. However, it is a verification of Einstein's prediction made decades ago.

One of the greatest explosions in general relativity has been in the area of cosmology. In 1965, two physicists, Robert Wilson and Arno Penzias, detected the faint microwave radiation from outer space with their Bell Laboratory Horn Radio Telescope in New Jersey. The two physicists, unaware of the pioneering work of Gamow and his students, accidentally picked up this cosmic radiation from the big bang without realizing it. (According to legend, they thought they were picking up interference from the bird droppings that littered their radio telescope. Later, Princeton physicist R. H. Dicke correctly identified this radiation as Gamow's microwave background radiation.) Penzias and Wilson were awarded the Nobel Prize for their pioneering work. Since then, the COBE (Cosmic Background Explorer) satellite, launched in 1989, has given us the most detailed picture of this cosmic microwave background radiation, which is remarkably smooth. When physicists led by George Smoot of the University of California at Berkeley carefully analyzed any slight ripples in this smooth background, they produced a remarkable photograph of the background radiation when the universe was only about 400,000 years old. The media mistakenly called this picture the "face of God." (This photograph is not the face of God, but it is a "baby picture" of the big bang.)

What is interesting about the picture is that the ripples probably correspond to tiny quantum fluctuations in the big bang. According to the uncertainty principle, the big bang could not have been a perfectly smooth explosion, since quantum effects must have produced ripples of a certain size. This, in fact, was precisely what the Berkeley group found. (In fact, if they had not found these ripples, it would have been a great setback for the uncertainty principle.) These ripples not only showed that the uncertainty principle applied to the birth of the universe, but also gave scientists a plausible mechanism for the creation of our "lumpy universe." When we look around us, we see that the galaxies are found in clusters, thereby giving the universe a rough texture. This lumpiness can possibly be easily explained as the ripples from the original big bang, which have been stretched as the universe expanded. Hence, when we see the clusters of galaxies in the heavens, we may be peering into the original ripples of the big bang left by the uncertainty principle.

But perhaps the most spectacular rediscovery of Einstein's work comes in the form of "dark energy." As we saw earlier, he introduced the concept of the cosmological constant (or the energy of the vacuum) in 1917 in order to prevent the universe from expanding. (We recall that there are only two possible terms allowed by general covariance, the Ricci curvature and the volume of space-time, so the cosmological constant term cannot be easily dismissed.) He later called it his greatest blunder when Edwin Hubble showed that the universe is in fact expanding. Results found in 2000, however, reveal that Einstein was probably right after all: the cosmological constant not only exists, but dark energy probably makes up the largest source of matter/energy in the entire universe. By analyzing supernovae in distant galaxies, astronomers have been able to calculate the rate of expansion of the universe over billions of years. To their surprise, they found that the expansion of the universe, instead of slowing down as most had thought, is actually speeding up. Our universe is in a runaway mode and will eventually expand forever. Thus, we can now predict how our universe will die.

Previously, some cosmologists believed that there might be enough matter in the universe to reverse the cosmic expansion, so that the universe might eventually contract and a blue shift would be seen in outer space. (Physicist Stephen Hawking even believed that time might reverse itself as the universe contracted and history might repeat itself in a backward fashion. This would mean that people would turn younger and jump into their mother's womb, that people would dive backward from a swimming pool and land dry on the diving board, and frying eggs would leap into their unbroken shells. Hawking, however, has since admitted he made a mistake.) Eventually, the universe would implode on itself, creating the enormous heat of a "big crunch." Others even speculated that the universe may then undergo another big bang, thereby creating an oscillating universe.

However, all this has now been ruled out with the experimental result that the expansion of the universe is accelerating. The simplest explanation that seems to fit the data is to assume that there is an enormous amount of dark energy pervading the universe which acts like antigravity, pushing the galaxies apart. The greater the universe becomes, the more vacuum energy there is, which in turn pushes the galaxies even farther apart, creating an accelerating universe.

This seems to vindicate one version of the "inflationary universe" idea, first proposed by MIT physicist Alan Guth, which is a modification of the original big bang theory of Friedmann and Lemaître. Roughly, in the inflationary picture there are two phases to the expansion. The first is a rapid, exponential expansion, when the universe was dominated by a large cosmological constant. Eventually, this exponential inflation terminates, and the expansion slows down to resemble the conventional expanding universe found by Friedmann and Lemaître. If correct, this means that the universe visible around us is just a pinpoint on a much larger space-time that represents the true universe. Recent experiments with balloons high in the atmosphere have also given credible evidence of inflation by showing that the universe seems to be approximately flat, which indicates how big it really is. We are like ants sitting on a huge balloon, thinking that our universe is flat only because we are so small.

Dark energy also forces us to reappraise our true role and position in the universe. It was Copernicus who showed that there was nothing special about the position of humans in the solar system. The existence of dark matter shows that there is nothing special about the atoms that make up our world, since 90% of the matter in the universe is made of mysterious dark matter. Now, the result from the cosmological constant indicates that dark energy dwarfs dark matter, which in turn dwarfs the energy of the stars and galaxies. The cosmological constant, once reluctantly introduced by Einstein to stabilize the universe, is probably by far the largest source of energy in the universe. (In 2003, the WMAP satellite verified that 4% of the universe's matter and energy is found in ordinary atoms, 23% in some form of unknown dark matter, and 73% of it in dark energy.)

Another strange prediction of general relativity is the black hole, which was considered science fiction when Schwarzschild reintroduced the concept of dark stars back in 1916. However, the Hubble Space Telescope and the Very Large Array Radio Telescope have now verified the existence of over fifty black holes, mainly lurking in the heart of large galaxies. In fact, many astronomers now believe that perhaps half of all the trillions of galaxies in the heavens have black holes at their center.

Einstein realized the problem with identifying these exotic creatures: by definition, they are invisible since light itself cannot escape, and hence extremely difficult to see in nature. The Hubble Space Telescope, peering into the hearts of distant quasars and galaxies, has now taken spectacular photographs of the spinning disk surrounding the black holes located in the heart of distant galaxies, such as M-87 and NGC-4258. In fact, one can clock some of this matter revolving around the black hole at about a million miles per hour. The most detailed Hubble photographs show that there is a dot at the very center of the black hole, about a single light-year across, which is powerful enough to spin an entire galaxy about 100,000 light-years across. After years of speculation, it was finally shown in 2002 that there is a black hole lurking in our own backyard, the Milky Way galaxy, which weighs the same as about 2 million suns. Thus, our moon revolves around the earth, the earth revolves around the sun, and the sun revolves around a black hole.

According to the work of Mitchell and Laplace in the eighteenth century, the mass of a dark star or black hole is proportional to its radius. Thus, the black hole at the center of our galaxy is roughly a tenth of the radius of the orbit of Mercury. It is astonishing that an object that small can affect the dynamics of our entire galaxy. In 2001, astronomers using the Einstein lens effect announced that a wandering black hole was discovered moving within the Milky Way galaxy. As the black hole moved, it distorted the surrounding starlight. By tracing the movement of this light distortion, astronomers could calculate its trajectory across the heavens. (Any wandering black hole approaching the earth could have catastrophic consequences. It would eat up the entire solar system and not even burp.)

In 1963, research in black holes received a boost when New Zealand mathematician Roy Kerr generalized Schwarzschild's black hole to include spinning black holes. Since everything in the universe seems to be spinning, and because objects spin faster when they collapse, it was natural to assume that any realistic black hole would be spinning at a fantastic rate. Much to everyone's surprise, Kerr found an exact solution of Einstein's equations in which a star collapsed into a spinning ring. Gravity would try to collapse the ring, but centrifugal effects could become sufficiently strong to counteract gravity, and the spinning ring would be stable. What most puzzled relativists was that if you fell through the ring, you would not be crushed to death. Gravity was actually large but finite at the center, so you could in principle fall straight through the ring, into another universe. A journey through the Einstein-Rosen bridge would not necessarily be a lethal one. If the ring were large enough, one might enter the parallel universe safely.

Physicists immediately began to pick apart what might happen if you fell into a Kerr black hole. An encounter with such a black hole would certainly be an unforgettable experience. In principle, it might give us a shortcut to the stars, transporting us instantly into another part of the galaxy, or perhaps another universe entirely. As you approached the Kerr black hole, you would pass through the event horizon so you would never be able to go back to where you started (unless there was another Kerr black hole that connected the parallel universe back to our universe, making a roundtrip possible). Also, there were problems with stability. One could show that if you fell through the Einstein-Rosen bridge, the distortions of space-time that you created might force the Kerr black hole to close up, making a complete journey through the bridge impossible.

As strange as the idea of a Kerr black hole was, acting as a gateway or portal between two universes, it could not be dismissed on physical grounds because black holes are indeed spinning very rapidly. However, it soon became apparent that these black holes not only connected two distant points in space, but also connected two times as well, acting as time machines.

When Gödel found the first time travel solution of Einstein's equations in 1949, it was considered a novelty, an isolated aberration of the equations. Since then, however, scores of time travel solutions have now been discovered in Einstein's equations. For example, it was discovered that an old solution, discovered by W. J. van Stockum in 1936, actually allowed for time travel. The van Stockum solution consisted of an infinite cylinder spinning rapidly around its axis, like the spinning pole found in old barbershops. If you journeyed around the spinning cylinder, then you might be able to return to the original spot before you left, much like the Gödel solution of 1949. Although this solution is intriguing, the problem is that the cylinder has to be infinite in length. A finite spinning cylinder will apparently not work. In principle, therefore, both the Gödel and the van Stockum solution can be ruled out on physical grounds.

In 1988, Kip Thorne and his colleagues at Caltech found yet another solution of Einstein's equations that admits time travel via a wormhole. They were able to solve the problem of the one-way trip through the event horizon by showing that a new type of wormhole was completely transversable. In fact, they have calculated that a trip through such a time machine may be as comfortable as a plane ride.

The key to all these time machines is the matter or energy that warps space-time onto itself. To bend time into a pretzel, one needs a fantastic amount of energy, far beyond anything known to modern science. For the Thorne time machine, one needs negative matter or negative energy. No one has ever seen negative matter before. In fact, if you had a piece of it in your hand, it would fall up, not down. Searches for negative matter have proved fruitless. If any existed on the earth billions of years ago, it would have fallen up into outer space, to be lost forever. Negative energy, on the other hand, actually exists in the form of the Casimir effect. If we take two neutral parallel metal plates, we know that they are uncharged and hence are not attracted or repelled toward each other. They should remain at rest. However, in 1948 Henrik Casimir demonstrated a curious quantum effect, demonstrating that the two parallel plates will actually attract each other by a small but nonzero force, which has actually been measured in the laboratory.

Thus a Thorne time machine can be built as follows: Take two sets of parallel metal plates. Because of the Casimir effect, the region between each set of plates will have negative energy. According to Einstein's theory, the presence of negative energy will open up tiny holes or bubbles in space-time (smaller than a subatomic particle) inside this region. Now assume, for the sake of argument, that an advanced civilization far ahead of ours can somehow manipulate these holes, grab one from each pair of plates, and then stretch them until a long tube or wormhole connects the two sets of plates. (Linking these two sets of parallel plates with a wormhole is far beyond anything possible with today's technology.) Now send one pair of plates on a rocket that is traveling near the speed of light, so that time slows down aboard the rocket ship. As we discussed earlier, clocks on the rocket run slower than clocks on Earth. If you jump into the hole within the parallel plates sitting on Earth, you will be sucked through the wormhole connecting the two plates and find yourself on the rocket back in the past, at a different point in space and time.

Since then, the field of time machines (or more properly "closed time like curves") has become a lively area of physics, with scores of papers published with different designs, all of them based on Einstein's theory. Not every physicist has been amused, though. Hawking, for one, did not like the idea of time travel. He said, tongue-in-cheek, that if time travel were possible, we would be flooded with tourists from the future, which we don't see. If time machines were commonplace, then history would be impossible to write, changing anytime someone spun the dial of their time machine. Hawking has declared that he wants to make the world safe for historians. However, in T. H. White's *The Once and Future King*, there is a society of ants that obeys the dictum, "Everything not forbidden is compulsory." Physicists take this law to heart, so Hawking was forced to postulate the "chronology protection conjecture," which bans time machines by fiat. (Hawking has since given up trying to prove this conjecture. He now maintains that time machines, although theoretically possible, are not practical.)

These time machines apparently obey the laws of physics, as we currently know them. The trick, of course, is to somehow access these tremendous energies (available only to "sufficiently advanced civilizations") and show that these wormholes are in fact stable against quantum corrections and don't explode or close up as soon as you enter one.

It should also be mentioned that time paradoxes (such as killing your parents before you are born) might be resolved with time machines. Because Einstein's theory is based on smooth, curved Riemann surfaces, we do not simply disappear when we enter the past and create a time paradox. There are two possible resolutions of time travel paradoxes. First, if the river of time can have whirlpools, then perhaps we simply fulfill the past when we enter the time machine. This means that time travel is possible, but we cannot alter the past, merely complete it. It was meant to be that we would enter the time machine. This view is held by Russian cosmologist Igor Novikov, who says, "We cannot send a time traveler back to the Garden of Eden to ask Eve not to pick the apple from the tree." Second, the river of time itself may fork into two rivers; that is, a parallel universe may open up. Thus, if you shoot your parents before you are born, you have only killed people who are genetically identical to your parents but are not really your parents at all. Your own parents indeed gave birth to you and made your body possible. What has happened is that you have jumped between our universe and another universe, so all time paradoxes are resolved.

But the theory closest to Einstein's heart was his unified field theory. Einstein remarked to Helen Dukas that perhaps in a hundred years, physicists will understand what he was doing. He was wrong. In less than fifty years, there has been a resurgence of interest in the unified field theory. The quest for unification, once derided by physicists as being hopelessly beyond reach, is perhaps now tantalizingly within our grasp. It dominates the agenda of almost every meeting of theoretical physicists.

After two thousand years of investigation into the properties of matter, ever since Democritus and fellow Greeks asked what the universe was made of, physics has produced two competing theories that are totally incompatible. The first is the quantum theory, which is incomparable in terms of describing the world of atoms and subatomic particles. The second is Einstein's general relativity, which has given us breathtaking theories of black holes and the expanding universe. The ultimate paradox is that these two theories are total opposites. They are based on different assumptions, different mathematics, and different physical pictures. The quantum theory is based on discrete packets of energy, called "quanta," and the dance of subatomic particles. The relativity theory, however, is based on smooth surfaces.

Physicists today have formulated the most advanced version of quantum physics, embodied in something called the "Standard Model," which can explain subatomic experimental data. It is, in some sense, the most successful theory in nature, able to describe the properties of three (the electromagnetic and the weak and strong nuclear forces) of the four fundamental forces. As successful as the Standard Model is, there are two glaring problems with it. First, it is supremely ugly, perhaps one of the ugliest theories ever proposed in science. The theory simply ties together the weak, strong, and electromagnetic forces by hand. It's like using Scotch tape to connect a whale, aardvark, and a giraffe together, and claiming that this is the supreme achievement of nature, the end product of millions of years of evolution. Up close, the Standard Model consists of a bewildering, motley collection of subatomic particles with strange names that do not make much sense, like quarks, Higgs bosons, Yang-Mills particles, W-bosons, gluons, and neutrinos. Worse, the Standard Model does not mention gravity at all. In fact, if one tries to graft gravity onto the Standard Model by hand, one finds that the theory blows up. It yields nonsense. All attempts for almost fifty years to graft the quantum theory and relativity together have proved fruitless. Given all its aesthetic defects, we conclude that the only thing going for the theory is that it is undeniably correct within its experimental domain. Clearly, what is needed is to go beyond the Standard Model, to re-examine the unification approach of Einstein.

After fifty years, the leading candidate for a theory of everything, one that can unify both the quantum theory and general relativity, is something called "superstring theory." In fact, it is the only game in town because all rival theories have been ruled out. As physicist Steven Weinberg said, "String theory has provided our first plausible candidate for a final theory." Weinberg believes that the maps that guided the ancient mariners all pointed to the existence of a fabled North Pole, though it would take centuries before Robert Peary actually set foot on it in 1909. Similarly, all the discoveries made in particle physics point to the existence of the "North Pole" of the universe, that is, a unified field theory. Superstring theory can absorb all the good features of the quantum theory and relativity in a surprisingly simple way. Superstring theory is based on the idea that subatomic particles can be viewed as notes on a vibrating string. Although Einstein compared matter to wood because of all its tangled properties and seemingly chaotic nature, superstring theory reduces matter to music. (Einstein, who was an excellent violinist, probably would have liked this.)

At one point in the 1950s, physicists despaired of making sense of subatomic particles because new ones were being discovered all the time. J. Robert Oppenheimer, in disgust, once said, "The Nobel Prize in Physics should be given to the physicist who does *not* discover a new particle that year." These subatomic particles were given so many strange Greek names that Enrico Fermi said, "If I had known that there would be so many particles with Greek names, I would have become a botanist rather than a physicist." But in string theory, if one had a super microscope and could peer directly into an electron, one would find not a point particle, but a vibrating string. When the superstring vibrates in a different mode or note, it changes into a different subatomic particle, like a photon or a neutrino. In this picture, the subatomic particles that we see in nature can be viewed as the lowest octave of the superstring. Thus, the blizzard of subatomic particles discovered over the decades are simply notes on this superstring. The laws of chemistry, which seem so confusing and arbitrary, are the melodies played out on superstrings. The universe itself is a symphony of strings. And the laws of physics are nothing but harmonies of the superstring.

Superstring theory can also encompass all of Einstein's work on relativity. As the string moves in space-time, it forces the surrounding space around it to curve, precisely as Einstein had predicted back in 1915. In fact, superstring theory is inconsistent unless it can move in a space-time consistent with general relativity. As physicist Edward Witten has said, even if Einstein had never discovered the theory of general relativity, it might have been rediscovered via the string theory. Witten says, "String theory is extremely attractive because gravity is forced upon us. All known consistent string theories include gravity, so while gravity is impossible in quantum field theory as we have known it, it's obligatory in string theory."

However, string theory makes some other quite surprising predictions. Strings can only consistently move in ten dimensions (one dimension of time and nine dimensions of space). In fact, string theory is the only theory which fixes the dimensionality of its own space-time. Like the Kaluza-Klein theory of 1921, it can unify gravity with electromagnetism by assuming that higher dimensions can vibrate, creating forces that can spread throughout three dimensions like light. (If we add an eleventh dimension, then string theory allows for the possibility of membranes vibrating in hyperspace. This is called "M-theory," which can absorb string theory and provide new insights into the theory from the vantage point of the eleventh dimension.)

What would Einstein think of superstring theory if he were alive today? The physicist David Gross said, "Einstein would have been pleased with this, at least with the goal, if not the real-

ization. . . . He would have liked the fact that there is an underlying geometrical principle—which unfortunately, we don't
really understand." The essence of Einstein's unified field theory, as we saw, was to create matter (wood) out of geometry
(marble). Gross commented on this: "To build matter itself out
of geometry—that in a sense is what string theory does. . . . [It's]
a theory of gravity in which particles of matter as well as the
other forces of nature emerge in the same way that gravity
emerges from geometry." It is instructive to go back to Einstein's
early work on the unified field theory, from the vantage point of
string theory. The key to Einstein's genius was that he was able
to isolate the key symmetries of the universe that unify the laws
of nature. The symmetry that unifies space and time is the
Lorentz transformation, or rotations in four dimensions. The
symmetry behind gravity is general covariance, or arbitrary
coordinate transformations of space-time.

However, on Einstein's third try at a great unifying theory, he
failed, mainly because he lacked the symmetry that would unite
gravity and light, or unite marble (geometry) with wood (matter). He, of course, was acutely aware that he lacked a fundamental principle that would guide him through the thicket of
tensor calculus. He once wrote, "I believe that in order to make
real progress one must again ferret out some general principle
from nature."

But that is precisely what the superstring provides. The symmetry underlying the superstring is called "supersymmetry," a
strange and beautiful symmetry that unifies matter with forces.
As mentioned earlier, subatomic particles have a property called
"spin," acting as if they were spinning tops. The electron, proton, neutron, and quarks that make up the matter in the universes all have spin ½ and they are called "fermions," named after
Enrico Fermi, who explored the properties of particles with half-
integral spin. The quanta of forces, however, are based on electromagnetism (with spin 1) and gravitation (with spin 2).

Notice that they have integral spin, and are called "bosons" (after the work of Bose and Einstein). The key point is that in general, matter (wood) is made of fermions with half-integral spin, while forces (marble) are made of bosons with integral spin. *Supersymmetry unifies fermions and bosons.* This is the essential point, that supersymmetry allows for a unification of wood and marble, as Einstein wished. In fact, supersymmetry allows for a new type of geometry that has even surprised the mathematicians, called "superspace," which makes possible "supermarble." In this new approach, we must generalize the old dimensions of space and time to include new fermionic dimensions, which then allows us to create a "superforce" out of which all forces originated at the instant of creation.

Thus, some physicists have speculated that one must generalize Einstein's original principle of general covariance to read: *the equations of physics must be super covariant* (i.e., maintain the same form after a super covariant transformation).

Superstring theory allows us to reanalyze Einstein's old work on the unified field theory, but in an entirely new light. When we begin to analyze the solutions to the superstring equations, we encounter many of the bizarre spaces that Einstein pioneered back in the 1920s and 1930s. As we saw earlier, he was working with generalizations of Riemannian space, which today can correspond to some spaces found in string theory. Einstein was looking at these bizarre spaces one after the other, in agonizing fashion (including complex spaces, spaces with "torsion," "twisted spaces," "antisymmetric spaces," etc.), but he got lost because he lacked any guiding physical principle or picture to lead him out of the tangle of mathematics. This is where supersymmetry comes in—it acts as an organizing principle that allows us to analyze many of these spaces from a different perspective.

But is supersymmetry the symmetry that eluded Einstein for the last three decades of his life? The key to Einstein's unified field theory is that it was to be made of pure marble, that is, pure geometry. The ugly "wood" that infested his original relativity theory was to be absorbed into geometry. Supersymmetry might hold the key to a theory of pure marble. In this theory, one can introduce something called "superspace," in which space itself becomes supersymmetrized. In other words, there is the possibility that the *final unified field theory will be made of "supermarble," out of a new "supergeometry."*

Physicists now believe that at the instant of the big bang, all the symmetries of the world were unified, as Einstein believed. The four forces we see in nature (gravity, electromagnetism, and the strong and weak nuclear force) were unified into a single "superforce" at the instant of creation, and only later broke apart as the universe cooled. Einstein's quest for the unified field theory seemed impossible, only because today we see the four forces of the world horribly broken into four pieces. If we can turn back the clock 13.7 billion years, to the original big bang, we would see the cosmic unity of the universe displayed in full glory, as Einstein imagined.

Witten claims that string theory will one day dominate physics the same way that quantum mechanics dominated physics for the past half-century. However, there are still many formidable obstacles. The critics of the theory point out some of its weak spots. First, it is impossible to test directly. Since superstring theory is a theory of the universe, the only way to test it is to re-create the big bang, that is, create energies in an atom smasher that approximate the beginning of the universe. To do this would require an atom smasher the size of a galaxy. This is out of the question, even for an advanced civilization. However, most physics is done indirectly, so there are high hopes that the Large Hadron Collider (LHC) to be built outside of Geneva, Switzerland, will have enough energy to probe the theory. The LHC, when it is turned on in the near future, will accelerate protons to trillions of electron volts, sufficient to smash atoms apart. When examining the debris of such fantastic collisions, physicists hope to find a new kind of particle, the super-particle or "sparticle," which would represent a higher resonance or octave of the superstring.

There is even some speculation that dark matter may be made of sparticles. For example, the partner of the photon, called the "photino," is neutral in charge, stable, and has mass. If the universe were filled with a gas of photinos, we would not be able to see it, but it would act very much like dark matter. One day, if we ever identify the true nature of dark matter, it may provide an indirect proof of superstring theory.

Yet another way to test the theory indirectly is to analyze gravity waves from the big bang. When the LISA gravity wave detectors are launched into space in the next decade, they may eventually pick up gravity waves emitted one-trillionth of a second after the instant of creation. If these agree with predictions made from the string theory, the data might confirm the theory once and for all.

M-theory may also explain some of the puzzles that surround the old Kaluza-Klein universe. Recall that one serious objection to the Kaluza-Klein universe was that these higher dimensions could not be seen in the laboratory, and in fact must be much smaller than an atom (otherwise, atoms would float into these higher dimensions). But M-theory gives us a possible solution to this by assuming that our universe itself is a membrane floating in an infinite eleven-dimensional hyperspace. Thus, sub-atomic particles and atoms would be confined to our membrane (our universe), but gravity, being a distortion of hyperspace, can flow freely between universes.

This hypothesis, as strange as it may seem, can be tested. Ever since Isaac Newton, physicists have known that gravity decreases as the inverse square of the distance. In four spatial dimensions, gravity should decrease as the inverse cube of the distance. Thus, by measuring tiny deviations from a perfect inverse square law, one may detect the presence of other universes. Recently, it was conjectured that if there is a parallel universe only a millimeter away from our universe, it might be compatible with Newtonian gravity and also might be detectable with the LHC. This in turn has created a certain amount of excitement among physicists, realizing that one aspect of superstring theory might be testable soon, either by looking for sparticles or by looking for parallel universes a millimeter from ours.

These parallel universes might provide yet another explanation for dark matter. If there is a parallel universe nearby, we will not be able to see it or feel it (since matter is confined to our membrane universe) but we would be able to feel its gravity (which can travel between universes). To us, this would appear as if invisible space had some form of gravity, much like dark matter. In fact, some superstring theorists have speculated that perhaps dark matter can be explained as the gravity produced by a nearby parallel universe.

But the real problem of proving the correctness of superstring theory is not experiment. We don't have to build gigantic atom smashers or space satellites to verify the theory. The real problem is purely theoretical: if we are smart enough to completely solve the theory, we should be able to find all its solutions, which should include our universe, with its stars, galaxies, planets, and people. So far, no one on Earth is smart enough to completely solve these equations. Perhaps tomorrow, or perhaps decades from now, someone may announce that they have completely solved the theory. At that time, we will be able to tell whether it is a theory of everything, or a theory of nothing. Because string theory is so precise, without any adjustable parameters, there is nothing in between.

Will superstring theory or M-theory allow us to unify the laws of nature into a simple, coherent whole, as Einstein once said? At this point, it is too early to say. We are reminded of Einstein's words: "The creative principle resides in mathematics. In a certain sense, therefore, I hold it true that pure thought can grasp reality, as the ancients dreamed." Perhaps a young reader of this book will be inspired by this quest for a unification of all physical forces to complete this program.

So how should we re-evaluate Einstein's true legacy? Instead of saying that he should have gone fishing after 1925, perhaps a more fitting tribute might be as follows: *All physical knowledge at the fundamental level is contained in two pillars of physics, general relativity and the quantum theory. Einstein was the founder of the first, was the godfather of the second, and paved the way for the possible unification of both.*

Notes

PREFACE

Page

xi   "A pop icon on a par . . .": Brian, p. 436.

xiv   "In the remaining 30 years of his life . . .": Pais, *Einstein Lived Here*, p. 43.

CHAPTER 1. PHYSICS BEFORE EINSTEIN

3   "If A is success, I should say . . .": Pais, *Einstein Lived Here*, p. 152.

3   "Everyone who had real contact . . .": French, p. 171.

5   "tortured man, an extremely neurotic . . .": Cropper, p. 19.

8   "is the most profound and the most fruitful that physics . . .": *Ibid.*, p. 173.

8   "The idea of the *time* of magnetic action . . .": *Ibid.*, p. 163.

9   "We can scarcely avoid the conclusion . . .": *Ibid.*, p. 164.

CHAPTER 2. THE EARLY YEARS

14   "A sound skull is needed . . .": Brian, p. 3.

14   "It doesn't matter; . . .": Clark, p. 27.

14   "Classmates regarded Albert as a freak . . .": Brian, p. 3.

15   "Yes, that is true. . . .": Pais, *Subtle Is the Lord*, p. 38.

15   "It is, in fact, nothing short . . .": Cropper, p. 205.

15   "A wonder of such.nature . . .": Schilpp, p. 9.

15   "Through the reading of popular books . . .": *Ibid.*, p. 5.

16   "In all these years I never . . .": Pais, *Subtle Is the Lord*, p. 38.

16   "At the age of 12, . . .": Schilpp, p. 9.

17   "Soon the flight of his mathematical genius . . .": Sugimoto, p. 14.

17   "philosophical nonsense . . .": Brian, p. 7.

19   "I love the Swiss . . .": Clark, p. 65.

20   "Whoever approached him was captivated . . .": Folsing, p. 39.

20   "Many a young or elderly woman . . .": *Ibid.*, p. 44.

21   "Beloved sweetheart . . .": Brian, p. 12; Folsing, p. 42.

21   "a work which I read with breathless attention.": Schilpp, p. 15.

23   "such a principle resulted from a paradox upon which . . .": *Ibid.*, p. 53.

23   "All physical theories, their mathematical expression notwithstand-
      ing, . . .": Calaprice, p. 261.

23   "most fascinating subject at the time . . .": Clark, p. 55.

24   "You are a smart boy, Einstein, . . .": Pais, *Subtle Is the Lord*, p. 44;
      Brian, p. 31.

24   "You're enthusiastic, but hopeless at physics. . . .": Folsing, p. 57.

25   "something very great": Sugimoto, p. 19.

25   "I can go anywhere I want— . . .": Folsing, p. 71.

26   "My sweetheart has a very wicked tongue . . .": Brian, p. 31.

26   "This Miss Maric is causing me . . .": *Ibid.*, p. 47.

26   "By the time you're 30, she'll be an old witch.": *Ibid.*

27   "What's to become of her?": *Ibid.*, p. 25.

27   "who cannot gain entrance to a good family.": *Ibid.*

27   "I would have found [a job] . . .": Thorne, p. 69.

28   "By the mere existence of his stomach, . . .": Schilpp, p. 3.

28   "I am nothing but a burden to my relatives . . .": Pais, *Subtle Is the
      Lord*, p. 41.

30   "pissing ink": Brian, p. 69.

30   "worldly monastery.": *Ibid.*, p. 52.

30   "Many years later, he still recalled . . .": *Ibid.*, p. 53.

30   "sad fate did not permit [her father] . . .": *Ibid.*

30   "The door of the flat was open to allow the floor, . . .": Sugimoto, p.
      33.

31   "private lessons in mathematics and physics.": *Ibid.*, p. 31.

31   "These words of Epicurus applied to us: . . .": Brian, p. 55.

Chapter 3. Special Relativity and the "Miracle Year"

36   "The germ of the special relativity theory . . .": Folsing, p. 166.

37   "A storm broke loose in my mind.": Brian, p. 61.

37   "The solution came to me suddenly . . .": *Ibid.*

38   "I owe more to Maxwell than to anyone.": *Ibid.*, p. 152. Many biog-
      raphies trace Einstein's ideas back to the Michelson-Morley experi-

ment. But as Einstein himself made clear on several occasions, this experiment only peripherally affected his thinking. He was led to relativity theory via Maxwell's equations. The entire thrust of his original paper was to show that Maxwell's equations had a hidden symmetry revealed by his relativity theory, and that this should be elevated to a universal principle of physics.

38 "Thank you, I've completely solved the problem.": Folsing, p. 155; Pais, *Subtle Is the Lord*, p. 139.

39 "one of the most remarkable volumes in the whole . . .": Cropper, p. 206.

42 "The idea is amusing and enticing; . . .": Folsing, p. 196.

42 "for the time being . . .": *Ibid.*, p. 197.

42 "Imagine the audacity of such a step . . .": Brian, p. 71.

48 "From now on, space and time separately have vanished . . .": *Ibid.*, p. 72.

48 "The main thing is the content, . . .": *Ibid.*, p. 76.

48 "superfluous erudition": Cropper, p. 220.

48 "Since the mathematicians have attacked the relativity . . .": Clark, p. 159.

49 "might have remained stuck in its diapers.": Cropper, p. 220.

51 "As a student he was treated contemptuously by the professors . . .": Brian, p. 73.

51 "The festivities ended in the Hotel National, . . .": *Ibid.*, p. 75.

52 "He appeared in class in somewhat shabby attire, . . .": Cropper, p. 215.

54 Another paradox involves two objects, each shorter than the other. . . : Over the decades, scores of paradoxes have been introduced to illustrate the seemingly bizarre nature of special relativity. They usually involve two frames of reference traveling at different speeds that are making observations of the same object. The paradoxes arise because the observers in each frame see the same object in two entirely different ways. Almost all of them can be resolved using two observations. First, length contraction in one frame has to be balanced with time dilation in the other. If we forget to balance the distortion of space with the distortion of time, then the paradoxes arise. Second, paradoxes arise if we forget to bring the two frames together at the end. The final resolution of who is really younger or shorter can be achieved when we bring the two observers together in space and time

and compare them. Unless we bring them together, then it is possible to have two objects each shorter and younger than the other, which is impossible in Newtonian physics.

55   *"There once was a young lady named Bright . . ."*: Going faster than light in order to break the time barrier to go backward in time is not possible. As you approach the speed of light, your mass becomes nearly infinite, you are squeezed until you are almost infinitely thin, and time almost stops. Hence, the speed of light is the ultimate speed in the universe. However, I discuss possible loopholes to this later, when I write about wormholes and Einstein-Rosen bridges.

55   "mathematical physicists are unanimous . . .": Sugimoto, p. 44.

56   "The gentlemen in Berlin are gambling on me . . .": Cropper, p. 216.

56   "It seems that most members . . .": Folsing, p. 336.

58   "I live a very withdrawn life . . .": *Ibid.*, p. 332.

59   "She is all love for her great husband, . . .": Brian, p. 151.

Chapter 4. General Relativity and "the Happiest Thought of My Life"

64   "As an older friend, I must advise you . . .": Pais, *Subtle Is the Lord*, p. 239.

64   "I was sitting in a chair in the patent office . . .": *Ibid.*, p. 179; Folsing, p. 303.

66   "Do not Bodies act upon light at a Distance, . . .": Folsing, p. 435.

69   "When a blind beetle crawls over the surface . . .": Calaprice, p. 9.

70   "Grossman, you must help me or else . . .": Pais, *Subtle Is the Lord*, p. 212.

70   "Never in my life have I tormented myself . . .": Folsing, p. 315.

70   "Do not worry about your difficulties in mathematics; . . .": Calaprice, p. 252.

73   Mach's principle: More precisely, Mach's principle states that the inertia of an object, and hence its mass, is due to the presence of all the other masses in the universe, e.g., the distant stars. Mach restated an observation known as far back as Newton, that the surface of a spinning pail of water becomes depressed (due to centripetal forces). The faster the spin, the greater the depression of the surface. If all motions are relative, including rotations, then one can consider the pail to be at rest and all the distant stars to rotate around it. Thus,

reasoned Mach, it was the rotation of the distant stars that caused the stationary pail's surface to be depressed. Thus, the presence of the distant stars determines the inertial properties of the pail of water, including its mass. Einstein modified this law to mean that the gravitational field is uniquely determined by the distribution of masses in the universe.

74   "If everything fails, I'll pay for the thing . . .": Folsing, p. 320.

75   Einstein had dropped the Ricci curvature . . .: General covariance means that the equations retain the same form after a change of coordinates (today this is called a "gauge transformation"). Einstein did not appreciate in 1912 that this meant that the physical predictions of his theory also remained the same after a change of coordinates. Thus, in 1912 he found, to his horror, that his theory gave an infinite number of solutions for the gravitational field surrounding the sun. But three years later, he suddenly realized that all of these solutions described the same physical system, i.e., the sun. Thus, the Ricci curvature was a perfectly well-defined mathematical object that could uniquely describe the gravitational field around a star, according to Mach's principle.

76   "For some days, I was beyond myself with excitement . . .": Folsing, p. 374.

76   "Imagine my joy over the practicability . . .": *Ibid.*, p. 373.

76   "Hardly anyone who has truly understood . . .": *Ibid.*, p. 372.

78   "Russian hordes allied with Mongols and Negroes unleashed against the white race.": Brian, p. 89.

78   "The German Army . . .": Sugimoto, p. 51.

79   "Unbelievable what Europe has unleashed in its folly.": Folsing, p. 343.

79   The war and the great mental effort necessary . . .: The chaos caused by World War I almost closed the University of Berlin when students seized control of the campus and the rector. Faculty members immediately called Einstein to help negotiate their release. Einstein in turn called physicist Max Born to help make the perilous journey to negotiate with the students. Born would later write that they traveled "in the Bavarian Quarter through streets full of wild-looking and shouting youths with red badges . . . . Einstein was well known to be politically left wing, if not 'red', and would be an ideal person to help negotiate with the students" (Brian, p. 97). Einstein was recognized by the students, who then

gave him their demands. They agreed to let their prisoners go free if the newly elected Social Democratic president, Friedrich Ebert, consented. Einstein and Born then made the journey to the palace of the Reichschancellor and pleaded with the president, who then agreed to authorize the release of the prisoners. Born recalled later, "We left the palace in the Reichschancellor in the highest of spirits, with the feeling of having taken part in a historic event, and hoping indeed that the time of Prussian arrogance was finished, that it was all over with the Junkers, the hegemony of the aristocrats, the cliques of officials, and the military, that now the German democracy was victorious." Einstein and Born, two theoretical physicists interested in the secrets of the atom and the universe, had apparently found a more practical application for their talents: saving their university.

CHAPTER 5. THE NEW COPERNICUS

82 "Dear Mother—Good news today . . .": Sugimoto, p. 57.
82 "If he had *really* understood . . .": Calaprice, p. 97.
82 "There was an atmosphere of tense interest . . .": Parker, p. 124.
82 "After careful study of the plates . . .": *Ibid.*
83 "one of the greatest achievements in the history of human thought. . . .": Clark, p. 290; Parker, p. 124.
83 "There's a rumor . . .": Parker, p. 126.
83 "Don't be modest Eddington . . .": *Ibid.*
83 "Revolution in Science—New Theory of the Universe— . . .": Folsing, p. 445.
83 "All England is talking . . .": *Ibid.*
84 "Today in Germany I am called a German man of science, . . .": *Ibid.*, p. 451.
84 "At present, every coachman and every waiter . . .": *Ibid.*, p. 343.
84 "Since the flood of newspaper articles . . .": Cropper, p. 217.
85 "This world is a curious madhouse . . .": *Ibid.*, p. 217.
85 "I feel now something like a whore. . . .": Brian, p. 106.
85 "seem to have been seized with something like intellectual panic . . .": *Ibid.*, p. 102.
85 "The supposed astronomical proofs . . .": *Ibid.*, p. 101.
85 "I have read various articles on the fourth dimension, . . .": *Ibid.*, p. 102.

86   "cross-eyed physics . . . utterly mad . . .": *Ibid.*, p. 103.

86   "A new scientific truth does not as a rule prevail . . .": Folsing, p. 199.

86   "Great spirits have always encountered violent opposition . . .": Pais, *Einstein Lived Here*, p. 219.

86   "could have been predicted from the start— . . .": Sugimoto, p. 66.

87   "we should not drive away such a man . . .": Brian, p. 113.

87   He had finally rediscovered his Jewish roots. . . .: It should be pointed out that his Zionist colleagues often feared that Einstein, famous for speaking his mind, would say things they disapproved of. Einstein, for example, once thought that the Jewish homeland should be in Peru, stressing that no one should be unnecessarily displaced if Jews settled there. He often stated that friendship and mutual respect between the Jewish and Arab people were absolutely important factors in any successful attempt to create a Jewish state in the Middle East. He once wrote, "I should much rather see a reasonable agreement with the Arabs based on living together in peace than the creation of a Jewish state" (Calaprice, p. 135).

87   "making me conscious of my Jewish soul . . .": Brian, p. 120.

88   "It's like the Barnum circus!": *Ibid.*, p. 121.

88   "The ladies of New York . . .": Sugimoto, p. 74.

88   A mob of eight thousand squeezed . . .: Brian, p. 123.

88   "from possibly serious injury only by strenuous efforts . . .": *Ibid.*, p. 130.

89   "It was the first time in my life . . .": Pais, *Einstein Lived Here*, p. 154.

89   "Not until I was in America did I discover . . .": Folsing, p. 505.

89   "If your theories are sound, I understand . . .": Brian, p. 131.

89   "He has become the great fashion. . . .": Pais, *Einstein Lived Here*, p. 152.

90   "If a German were to discover a cure for cancer . . .": Sugimoto, p. 63.

90   Previously, he had even advocated killing Rathenau.: *Ibid.*, p. 64.

90   "it was a patriotic duty to shoot . . .": Clark, p. 360

90   Once, a mentally unbalanced Russian immigrant, . . .: Brian, p. 150.

91   "Life is like riding a bicycle. . . .": *Ibid.*, p. 146.

91   "He spent all his time . . .": Brian, p. 144.

92   By the 1920s and 1930s, Einstein had emerged as a giant . . .: Einstein, a lion of German society, was constantly surrounded by wealthy matrons who clamored to hear his wit and wisdom, many of whom

would donate generously to his favorite causes and charities. Some of them would occasionally send their personal limousine to pick Einstein up at his summer house in Caputh, to escort him to a fundraiser or concert. Inevitably, rumors would spread about alleged affairs. If one tracks the source of these rumors, one finds that they come mainly from the recollections of the maid at the summer house, Herta Waldow, who sold her story to the press. She had no proof, however, of any extramarital affairs and admitted that these society women would invariably personally give chocolates to Elsa when picking up her husband to quell any suspicions of impropriety. Furthermore, Konrad Wachsmann, an architect who helped to design the Caputh summer house, observed the Einstein household and concluded that these liaisons were perfectly harmless. He believed that they were "almost without exception" platonic in nature, that Einstein was never unfaithful to Elsa with these women.

92   "gentle, warm, motherly, . . .": Cropper, p. 217.

92   "He ate with everybody, . . .": Pais, *Einstein Lived Here*, p. 184.

93   "The people applaud me . . .": Sugimoto, p. 122.

93   "I could have imagined . . .": Brian, p. 205.

93   "It was interesting to see them together— . . .": Calaprice, p. 336.

94   "Is not all of philosophy as if written in honey? . . .": Pais, *Subtle Is the Lord*, p. 318.

94   "The world, considered from the physical aspect, . . .": Pais, *Einstein Lived Here*, p. 186.

94   "Morality is of the highest importance— . . .": Calaprice, p. 293.

94   "Science without religion is lame, . . .": Pais, *Einstein Lived Here*, p. 122.

94   "The most beautiful and deepest experience . . .": *Ibid.*, p. 119.

94   "If something is in me which can be called religious, . . .": Sugimoto, p. 113.

94   "I'm not an atheist . . .": Brian, p. 186.

Chapter 6. The Big Bang and Black Holes

96   "If the matter was evenly . . .": Misner et al., p. 756.

101   "My husband does that . . .": Croswell, p. 35.

105   "There should be a law of Nature . . .": Thorne, p. 210.

107   "there is not much hope . . .": Petters et al., p. 7.

107   "is of little value, but it makes the poor guy [Mandl] happy.": *Ibid*.

CHAPTER 7. UNIFICATION AND THE QUANTUM CHALLENGE

112   "They do not shake my strong feeling . . .": Pais, *Subtle Is the Lord*, p. 23.

114   "It is a masterful symphony": Parker, p. 209.

115   "no significance for physics.": Pais, *Subtle Is the Lord*, p. 343.

115   "The idea of achieving . . .": *Ibid.*, p. 330.

115   "The formal unity of your theory is startling.": *Ibid.*, p. 330.

118   "You may be amused to hear . . .": Pais, *Einstein Lived Here*, p. 179.

118   "It is not even wrong.": Cropper, p. 257.

118   "I do not mind . . .": *Ibid.*

118   "What you said . . .": *Ibid.*

118   "Some people have very sensitive . . .": *Ibid.*

119   "The more success . . .": Calaprice, p. 231.

121   "Like the dark lady who inspired Shakespeare's sonnets, . . .": Moore, p. 195.

123   This extra minus sign, argued Dirac, made possible . . .: Because matter prefers to tumble down to the lowest energy state, this meant that all electrons might fall into these negative energy states and the universe would collapse. To prevent this disaster, Dirac postulated that all negative energy states were already filled. A passing gamma ray might knock an electron out of its negative energy state, leaving a "hole" or bubble. This hole, predicted Dirac, would behave like an electron with positive charge, i.e., antimatter.

123   "The saddest chapter of modern physics . . .": Pais, *Inward Bound*, p. 348.

123   "I think that this discovery of anti-matter . . .": *Ibid.*, p. 360

125   "the motion of particles follows . . .": Folsing, p. 585.

127   "Quantum mechanics calls for a great deal of respect. . . .": *Ibid.*

127   "Heisenberg has laid a big quantum egg. . . .": Brian, p. 156.

127   "cobbler or employee in a gaming house": Ferris, p. 290.

127   "Physicists were beginning to . . .: Einstein most clearly presented his position on determinism and uncertainty as follows: "I am a determinist, compelled to act as if free will existed, because if I wish to live in a civilized society, I must act responsibly. I know philo-

sophically a murderer is not responsible for his crimes, but I prefer not to take tea with him. . . . I have no control, primarily those mysterious glands in which nature prepares the very essence of life. Henry Ford may call it his Inner Voice, Socrates referred to it as his daemon: each man explains in his own way the fact that the human will is not free. . . . Everything is determined, the beginning as well as the end, by forces over which we have no control. It is determined for the insect as well as for the star. Human beings, vegetables, or cosmic dust, we all dance to a mysterious time, intoned in the distance by an invisible player" (Brian, p. 185).

128   "If the last proof is sent away, then I will come.": Cropper, p. 244.

129   "To Bohr, this was a heavy blow. . . .": Folsing, p. 561.

129   "I am convinced that this theory . . .": *Ibid.*, p. 591.

130   "the greatest debate in intellectual history . . .": Brian, p. 306.

130   "I don't like it, . . .": Kaku, *Hyperspace*, p. 280.

132   "Does the moon exist . . .": *Ibid.*, p. 260.

132   "I have thought a hundred times as much about the quantum problems . . .": Calaprice, p. 260.

133   "spooky action-at-a-distance": Brian, p. 281.

133   "I was very happy that in that paper . . .": *Ibid.*

133   "We dropped everything; . . .": Folsing, p. 698.

134   "most successful physical theory of our period": Pais, *Einstein Lived Here*, p. 128.

CHAPTER 8. WAR, PEACE, AND E = MC²

138   "This means that I am opposed to the use of force . . .": Cropper, p. 226.

138   "The purpose of this publication is to oppose . . .": Sugimoto, p. 127.

139   "Turn around, you will never see it again.": Pais, *Einstein Lived Here*, p. 190.

139   "Under today's conditions, if I were a Belgian, . . .": Folsing, p. 675.

139   "The antimilitarists are falling on me . . .": *Ibid.*

140   "I had hoped to convince him . . .": Cropper, p. 271.

140   "People say that I get attacks of nervous weakness, . . .": Brian, p. 247.

140   "I failed to make myself understood . . .": Cropper, p. 271.

141   He could have been seriously hurt by this ferocious beating, . . .: Moore, p. 265.

142 "Princeton is a wonderful little spot . . .": Cropper, p. 226.

142 "large wastebasket . . . so I can throw . . .": Brian, p. 251.

142 Two Europeans, on a bet . . .: Parker, p. 17.

142 "grave heredity": Folsing, p. 672.

142 "I have seen it coming, . . .": *Ibid.*

143 "utterly ashen and shaken": Brian, p. 297.

143 "severed the strongest tie he had . . .": *Ibid.*

143 "I have got used extremely well . . .": Folsing, p. 699.

143 "It might be possible, and it is not even improbable, . . .": *Ibid.*, p. 707.

144 "All bombardments since . . .": *Ibid.*, p. 708.

144 "Assuming that it were possible to effect . . .": *Ibid.*

144 "as firing at birds in the dark, . . .": *Ibid.*, p. 709.

144 "the rays released . . . are in turn . . .": *Ibid.*, p. 708.

144 "the country that exploits it first . . .": *Ibid.*, p. 712.

145 "anyone who expects a source of power . . .": Pais, *Inward Bound*, p. 436.

146 "Oh, what fools we all have been!" Cropper, p. 340.

146 "do not justify the assumption . . .": Folsing, p. 710.

146 "Some recent work by E. Fermi and L. Szilard, . . .": *Ibid.*, p. 712.

147 "This requires action.": *Ibid.*

147 "I will have nothing . . .": Cropper, p. 342.

147 "I would rather walk naked . . .": *Ibid.*

148 "I wish very much that I could place . . .": Folsing, p. 714.

148 "In view of his radical background, . . .": *Ibid.*

148 "He felt very bad about being neglected. . . .": *Ibid.*, p. 715.

150 "said a new kind of bomb has been dropped on Japan. . . .": Brian, p. 344.

150 In 1946, Einstein made the cover . . .: In 1948, he helped draft his Message to the Intellectuals, which stated, "Man has not succeeded in developing political and economic forms of organization which would guarantee the peaceful coexistence of the nations of the world. We scientists, whose tragic destiny has been to help in making the methods of annihilation more gruesome and more effective, must consider it our solemn and transcendent duty to do all in our power to prevent these weapons from being used for the brutal purpose for which they were invented. What task could possibly be more important to us? What social aim could be closer to our hearts?"

(Sugimoto, p. 153).

He clarified his view on world government when he said, "The only salvation for civilization . . . lies in the creation of world government, with security of nations founded upon law. . . . As long as sovereign states continue to have separate armaments and armaments secrets, new world wars will be inevitable" (Folsing, p. 721).

151   "You are after big game. . . .": Brian, p. 350.

151   "I believe I am right. . . .": *Ibid.*, p. 359.

151   "Mathematical patterns like those of the painters or the poets . . .": Weinberg, p. 153.

152   "I have become a lonely old fellow. . . .": Brian, p. 331.

152   "I must seem like an ostrich . . .": Pais, *Subtle Is the Lord,* p. 465.

152   "I am generally regarded as a sort of petrified object . . .": *Ibid.*, p. 162.

152   "Oppenheimer made fun . . .": Brian, p. 377.

153   "This is not a jubilee book for me, . . .": Cropper, p. 223.

153   "Anything really new is invented . . .": *Ibid.*

153   "Nature shows us only . . .": Calaprice, p. 232.

153   "Subtle is the Lord, . . ." *Ibid.*, p. 241.

153   "I have second thoughts . . .": *Ibid.*

154   "From 1954 to the end of his life, . . .": Pais, *Inward Bound*, p. 585.

154   "We in the back . . .": Kaku, *Beyond Einstein*, p. 11.

154   "It was an uncanny encounter of two giants . . .": Cropper, p. 252.

157   "What I admired most about Michele was the fact . . .": Overbye, p. 377.

157   "It is tasteless to prolong life artificially. . . .": Calaprice, p. 63.

Chapter 9. Einstein's Prophetic Legacy

162   "The very study of the external world . . .": Crease and Mann, p. 67.

162   "Science cannot solve the ultimate mystery . . .": Barrow, p. 378.

162   This would be the acid test. . . .: More precisely, Bell advocated reexamining the old EPR experiment. In principle, one can measure the angles created by the axis of polarization of the pairs of electrons. By making a detailed analysis of the correlation between various angles of polarization between the two pairs of electrons, Bell was able to construct an inequality, called "Bell's inequality," con-

cerning these angles. If quantum mechanics were correct, then one set of relations would be satisfied. If quantum mechanics were incorrect, then another set of relations would be satisfied. Every time this experiment has been performed, the predictions of quantum mechanics prove to be correct.

165   "I think I can safely say . . .": Barrow, p. 144.

168   "At first sight, it looked artificial . . .": Petters et al., p. 155; *New York Times*, March 31, 1998.

170   "It's a bulls-eye!": *New York Times, Ibid.*

178   "We cannot send a time traveler . . .": Hawking et al., p. 85.

180   "String theory has provided our first plausible candidate . . .": Weinberg, p. 212.

180   "The Nobel Prize in Physics . . .": Kaku, *Beyond Einstein*, p. 67.

180   "If I had known . . .": *Ibid.*

181   "String theory is extremely attractive because gravity is forced upon us. . . .": Davies and Brown, p. 95. It should also be pointed out that the latest version of string theory is called "M-theory." String theory is defined in ten-dimensional space (with nine dimensions of space and one dimension of time). However, there are five self-consistent string theories that can be written in ten dimensions, which has puzzled theorists who would like a single candidate for a unified field theory, not five. Recently, Witten and his colleagues showed that all five theories are actually equivalent if one defines the theory in eleven-dimensional space (with ten dimensions of space and one dimension of time). In eleven dimensions, higher dimensional membranes can exist, and some speculate that our universe may be such a membrane. Although the introduction of M-theory has been a great advance for string theory, at present no one knows the precise equations for M-theory.

181   "Einstein would have been pleased with this, . . .": *Ibid.*, p. 150.

182   "I believe that in order to make real progress . . .": Pais, *Subtle Is the Lord*, p. 328.

186   "The creative principle resides . . .": Kaku, *Quantum Field Theory*, p. 699.

Bibliography

According to his will, Einstein donated all his manuscripts and letters in the Einstein Archives to Hebrew University in Jerusalem. Copies of the documents can be found at Princeton University and Boston University. *The Collected Papers of Albert Einstein* (vols. 1 through 5), edited by John Stachel, provides translations of this voluminous material.

Barrow, John D. *The Universe That Discovered Itself.* Oxford University Press, Oxford, 2000.

Bartusiak, Marcia. *Einstein's Unfinished Symphony.* Joseph Henry Press, Washington, D.C., 2000.

Bodanis, David. $E = mc^2$. Walker, New York, 2000.

Brian, Denis. *Einstein: A Life.* John Wiley and Sons, New York, 1996.

Calaprice, Alice, ed. *The Expanded Quotable Einstein.* Princeton University Press, Princeton, 2000.

Clark, Ronald. *Einstein: The Life and Times.* World Publishing, New York, 1971.

Crease, R., and Mann, C. C. *Second Creation.* Macmillan, New York, 1986.

Cropper, William H. *Great Physicists.* Oxford University Press, New York, 2001.

Croswell, Ken. *The Universe at Midnight.* Free Press, New York, 2001.

Davies, P. C. W., and Brown, Julian, eds. *Superstrings: A Theory of Everything?* Cambridge University Press, New York, 1988.

Einstein, Albert. *Ideas and Opinions.* Random House, New York, 1954.

Einstein, Albert. *The Meaning of Relativity.* Princeton University Press, Princeton, 1953.

Einstein, Albert. *Relativity: The Special and the General Theory.* Routledge, New York, 2001.

Einstein, Albert. *The World as I See It.* Kensington, New York, 2000.

Einstein, Albert, Lorentz, H. A., Weyl, H., and Minkowski, H. *The Principle of Relativity.* Dover, New York, 1952.

Ferris, Timothy. *Coming of Age in the Milky Way.* Anchor Books, New York, 1988.

Flückiger, Max. *Albert Einstein in Bern.* Paul Haupt, Bern, 1972.

Folsing, Albrecht. *Albert Einstein.* Penguin Books, New York, 1997.

Frank, Philip. *Einstein: His Life and His Thoughts.* Alfred A. Knopf, New York, 1949.

French, A. P., ed. *Einstein: A Centenary Volume.* Harvard University Press, Cambridge, 1979.

Gell-Mann, Murray. *The Quark and the Jaguar.* W. H. Freeman, San Francisco, 1994.

Goldsmith, Donald. *The Runaway Universe.* Perseus Books, Cambridge, Mass., 2000.

Hawking, Stephen, Thorne, Kip, Novikov, Igor, Ferris, Timothy, and Lightman, Alan. *The Future of Spacetime.* W. W. Norton, New York, 2002.

Highfield, Roger, and Carter, Paul. *The Private Lives of Albert Einstein.* St. Martin's, New York, 1993.

Hoffman, Banesh, and Dukas, Helen. *Albert Einstein, Creator and Rebel.* Penguin, New York, 1973.

Kaku, Michio. *Beyond Einstein.* Anchor Books, New York, 1995.

Kaku, Michio. *Hyperspace.* Anchor Books, New York, 1994.

Kaku, Michio. *Quantum Field Theory.* Oxford University Press, New York, 1993.

Kragh, Helge. *Quantum Generations.* Princeton University Press, Princeton, 1999.

Miller, Arthur I. *Einstein, Picasso.* Perseus Books, New York, 2001.

Misner, C. W., Thorne, K. S., and Wheller, J. A. *Gravitation.* W. H. Freeman, San Francisco, 1973.

Moore, Walter. *Schrödinger, Life and Thought.* Cambridge University Press, Cambridge, 1989.

Overbye, Dennis. *Einstein in Love: A Scientific Romance.* Viking, New York, 2000.

Pais, Abraham. *Einstein Lived Here: Essays for the Layman.* Oxford University Press, New York, 1994.

Pais, Abraham. *Inward Bound: Of Matter and Forces in the Physical World.* Oxford University Press, New York, 1986.

Pais, Abraham. *Subtle Is the Lord —: The Science and the Life of Albert*

Einstein. Oxford University Press, New York, 1982.

Parker, Barry. *Einstein's Brainchild: Relativity Made Relatively Easy.* Prometheus Books, Amherst, N.Y., 2000.

Petters, A. O., Levine, H., and Wambganss, J. *Singularity Theory and Gravitational Lensing.* Birkhauser, Boston, 2001.

Sayen, Jamie. *Einstein in America.* Crown Books, New York, 1985.

Schilpp, Paul. *Albert Einstein: Philosopher-Scientist.* Tudor, New York, 1951.

Seelig, Carl. *Albert Einstein.* Staples Press, London, 1956.

Silk, Joseph. *The Big Bang.* W. H. Freeman, San Francisco, 2001.

Stachel, John, ed. *The Collected Papers of Albert Einstein*, vols. 1 and 2. Princeton University Press, Princeton, 1989.

Stachel, John, ed. *Einstein's Miraculous Year.* Princeton University Press, Princeton, 1998.

Sugimoto, Kenji. *Albert Einstein: A Photographic Biography.* Schocken Books, New York, 1989.

Thorne, Kip S. *Black Holes and Time Warps: Einstein's Outrageous Legacy.* W. W. Norton, New York, 1994.

Trefil, James S. *The Moment of Creation.* Collier Books, New York, 1983.

Weinberg, Steven. *Dreams of a Final Theory.* Pantheon Books, New York, 1992.

Zackheim, Michele. *Einstein's Daughter.* Riverhead Books, New York, 1999.

Zee, A. *Einstein's Universe: Gravity at Work and Play.* Oxford University Press, New York, 1989.